Growing
Biodynamic
Crops

Growing Biodynamic Crops

Sowing, Cultivation and Rotation

Friedrich Sattler &
Eckard von Wistinghausen

Floris Books

Translated by A.R. Meuss

First published in German as part of *Der landwirtschaftlicher Betrieb: biologisch-dynamisch* in 1985 by Eugen Ulmer, Stuttgart
Second German edition published in 1989
First published in English as part of *Bio-Dynamic Farming Practice* by the Bio-Dynamic Agricultural Association in 1992

This excerpt published in 2014 by Floris Books in cooperation with the Biodynamic Association, Stroud

© 1989 Eugen Ulmer GmbH & Co, Stuttgart, Germany
Translation © A.R. Meuss 1992

All rights reserved. No part of this publication may be reproduced without the prior permission of Floris Books, 15 Harrison Gardens, Edinburgh
www.florisbooks.co.uk

British Library CIP data available
ISBN 978-178250-112-1
Printed in Great Britain
by CPI Group (UK) Ltd, Croydon

Contents

Foreword	9
1. The Nature of Plants	11
2. Factors Affecting Plant Growth	13
3. Land Use	19
4. Grassland	21
Cultivations	22
Manuring	23
Meadows	25
Pasture and pastures mown at intervals	26
Spring and autumn grazing technique 32;	
Watering 33; Cleansing cut 35	
Grassland weed control	35
Ploughing and reseeding of permanent grassland	38
Hedges and trees in grassland	42
5. Crops and Cropping	44
A look at the history	44
The sidereal system	46
Two and three-field system 47; Crop rotation 47	

6. Cropping Sequence 48
   Percentage areas 48; Fertilizer and fodder requirements 49; Soil-improving crops 49
   Economic and labour aspects 50
   Weed and pest control by use of cropping sequences 51
   Compatibilities 53
   Principles and terminology 55
   Terminology 55
   Preceding crop value and effect 56
   Cereals and grasses 58; Non-cereals 60
   Catch crops 61
   Manuring as part of the sequence 63
   Examples of cropping sequences 64

7. Seeds and Sowing Techniques 73
   Choice of variety 74
   Growing quality seed 75
   Sowing times 76; Choice of growing site 77; In-field selection 78; Seed baths 79
   External qualities of seed 81
   Size, weight, shape 81; Clean seed 82; Germinating power 82; Water content 83
   Sowing techniques 86
   Density 86; Depth 87; Distance between rows 88
   Seed bed preparation and sowing 89

8. Post Drilling Cultivations 91
   Available methods 94
   Cropping sequences 95; Mechanical measures 98; Thermal control 102
   A practical example from a farm 104
   Winter cereals 105; Spring cereals 107

9. Arable Crops — 109

Cereals — 109
Wheat 112; Rye 113; Barley 115; Oats 116; Smut and seed treatments 117

Row crops — 118
Potatoes 120; Sugar beet 133; Fodder beet, mangels and swedes 133; Beetroot 136; Carrots 137; White and red cabbage 141

Legumes — 143
Peas 144; Field beans 146; Vetches 147; Lentils 148; Lupins 149

Miscellaneous crops — 150
Winter rape 150; Flax 152; Mustard 154; Buckwheat 154; Stinging nettle 155

Fodder crops — 157
Annual clovers 162; Perennial clovers 164; Leys 169; Non-leguminous fodder crops 179; Catch crops 183

Growing fodder plants for seed — 192

Herbs — 197
Cultivation 204; Harvesting and processing 205; Growing peppermint 208; Herb fodder 209

Bibliography — 211

Index — 213

Abbreviations used

CU	Cattle Unit
kSE	kilo Starch Equivalent (1000 SE)
NEL	Net Energy Lactation
TGW	thousand grain weight

Conversion to US measures

Metric measures are used in this book. Within the text US equivalents are shown (US gallon and short ton, *not* imperial gallon or long ton). However, within tables this has not been possible for reasons of space. To convert from metric to US measure multiply by the following:

1 mm	0.039 in
1 cm	0.394 in
1 m	3.28 ft
1 m²	10.76 sq ft
1 ha	2.47 acre
1 m³	35.3 cu ft
1 g	0.0353 oz
1 kg	2.20 lb
1 t	1.10 tons
1 litre	0.264 gal
1 unit/ha	2.47 units/acre
1 kg/ha	0.892 lb/acre
1 t/ha	0.446 tons/acre
1 litre/ha	0.107 gal/acre

Foreword

Growing Biodynamic Crops originally formed part of *Bio-Dynamic Farming Practice*, the detailed manual on biodynamic agriculture published in German in 1989 and in English in 1992. It was a book that became an indespensible English language reference work for aspiring biodynamic farmers and a standard textbook for biodynamic students. Although times have moved on, the wisdom and practical know-how contained within it remains as valid as it ever was. Fritz Sattler managed a farm for 33 years in southern Germany. His lifetime's experience of both arable and livestock farming provides a rich source of information dating back to the pioneering years of biodynamic agriculture. The subject of crop husbandry dealt with in this edition is particularly well researched. Many aspects are of course not unique to biodynamic agriculture. Principles of grassland management, which are so carefully described in the first chapters, draw on the highly regarded work of researchers such as Andre Voisin and Sir John Stapledon, as well as recommendations arising from organic field trials and the farmer's own extensive experience. Embedded within the text however are countless references to and recommendations for biodynamic applications.

Biodynamic agriculture is practiced today in many corners of the world where it is recognized as being a particularly sustainable farming approach. Fine taste, healthy appearance and good keeping quality are characteristics recognizable in many foodstuffs produced under biodynamic conditions. They are also less prone to disease, more resilient and have greater vitality. These qualities – along with a farming approach that is strictly organic, encourages diversity and observes the law of return – make biodynamic agriculture an approach of choice for increasing numbers of people concerned for the future of our planet.

Central to the biodynamic approach is the application of unique life-enhancing preparations for treating the soil and growing plants. Biodynamic sprays created entirely from natural materials, are used to enhance plant sensitivity and regulate the way they develop and ripen. One of these, known as 'horn manure', serves to strengthen and sensitize plant roots to the mineral and nutrient capacities of the soil while a second, 'horn silica', regulates the plant's metabolism, strengthens its structure and enhances the qualitative ripening processes. There are also preparations based on six different medicinal plants – yarrow, chamomile, nettle, oak bark, dandelion and valerian. These are the biodynamic compost preparations which are used to effect a harmonious transformation of decaying materials into humus-rich compost.

Another aspect of biodynamic practice referred to concerns the timing of particular operations in relation to specific sun, moon and planetary constellations. The well-known planting calendar of Maria Thun is only obliquely referred to but is a widely used tool in biodynamic practice. According to her research the passage of the moon through the zodiac constellations encourages the growth of particular plant organs. Thus when the moon stands in front of Capricorn, Taurus and Virgo, the roots of plants are stimulated while Pisces, Cancer and Scorpio support leaf growth, Aquarius, Gemini and Libra that of flowers and Aries, Leo and Sagittarius promote fruiting.

The book was written by German farmers and therefore has a distinct continental flavour. In Britain there were in the past very few farms growing cereals on any scale and for many years biodynamic agriculture was mostly practiced on smaller farms focusing on livestock and vegetable production. This is perhaps not surprising given that much of western Britain is best suited to livestock farming and that until fairly recently there were no biodynamic farms in the eastern cereal growing regions. This is now changing and as more biodynamic arable farms develop the guidance provided in this manual will be of invaluable support.

Bernard Jarman, July 2014

1. The Nature of Plants

The plant world acts as a link between sun and star activities on the one hand and the earth with its physical matter and forces on the other.

The flowering plant grows vertically upwards from the ground and towards the sun (heliotropism). The root is orientated towards the centre of the earth (geotropism); lateral roots and fine capillary roots establish a close relationship to the mineral and salt elements of the earth.

Roots passively take up water and dissolved nutrients as part of the transpiration process that is initiated by the sun. They are also capable of selectively taking up water and minerals from the soil to meet requirements. This capacity is enhanced by the presence of

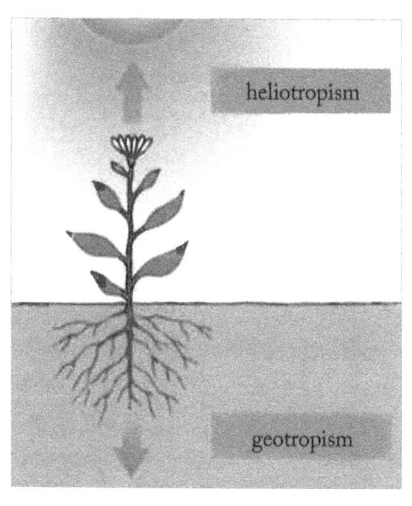

Figure 1. The plant between sun and earth.

humus, compost rich in nutrients, well rotted farmyard manure, a balanced mineral composition of the soil, and light effects that are largely mediated by the aerial parts of the plant.

With the aid of sunlight, the chlorophyll which is mainly present in green leaves enables plants to synthesize glucose and starch (carbohydrates) from water and atmospheric carbon dioxide. Oxygen is released in the process to provide the basis for all life on earth.

Green plants are therefore the productive element in the whole ecology. The life processes of humans, animals and lower life forms such as fungi and bacteria depend on the plant kingdom.

2. Factors Affecting Plant Growth

Plant growth is closely bound up with the environmental conditions of the biotope. Individual plant species have widely differing requirements. The growth factors
- heat,
- light and air,
- water, and
- minerals

must be available in a balance suited to the needs of the species if the plants are to thrive.

Moisture and heat are important if seeds are to germinate. The temperature range in which germination takes place varies enormously with individual plants. Many have to experience the shock of low temperatures in the winter soil if they are to achieve normal growth; if this is lacking, winter cereals will not form ears, for example. Heat generally promotes life processes, but in the absence of other factors such as water it may also harm plants. More sugar may be broken down in the respiratory process during extremely warm nights than has been synthesized by assimilation during the day.

The relationship of plants to light and air is a particularly close one. Examples of this are
- photosynthesis,
- germination (seeds needing either light or darkness to germinate),
- intensification of root growth in stronger light,
- phototropism, i.e. turning to the light,
- short- and long-day plants,

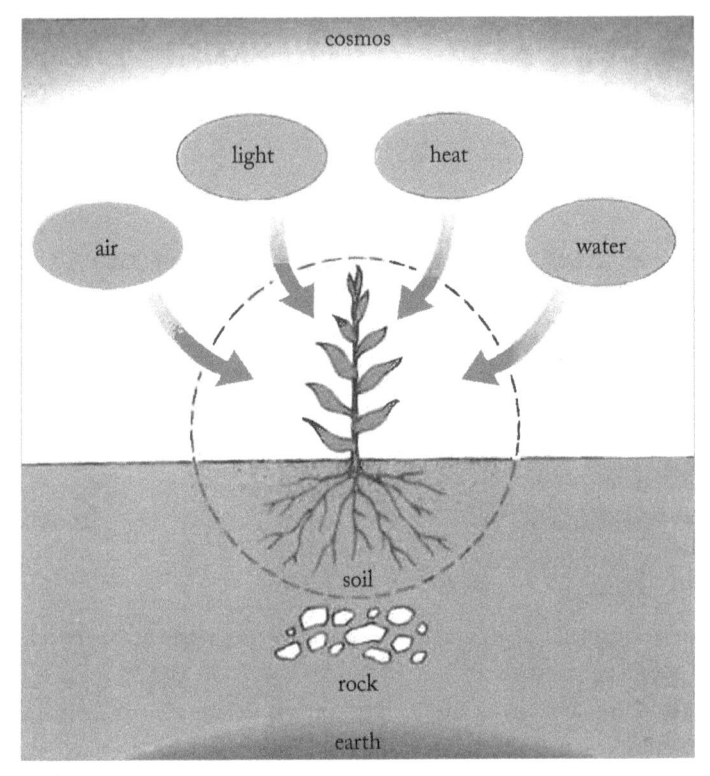

Figure 2. Factors influencing plant growth.

- damage due to lack of light (susceptibility to fungal attack and pests, little or no nutrient uptake through roots, and many other effects),
- damage due to bad atmospheric conditions (sulfur).

Air is the provider of carbon dioxide for assimilation and oxygen for respiration. It must also be present in the soil, for soil organisms and roots require much air (oxygen) to enable them to function. What is more, the nitrogen in the air is fixed by free-living bacteria (*Azotobacter, Bacillus amylobacter, Clostridium*, etc.), fungi and algae (blue-green algae: *Nostoc, Calothrix* mainly in rice-growing soils), and above all bacteria living in symbiosis with plants (nitrogen-fixing bacteria and leguminous plants); it is then available to plants as organically bound nitrogen.

Water is particularly important in plant life as it makes up about 80% of a plant's body. It is involved in all vital processes:
- as a solvent in all metabolic processes,
- in biocycles,
- in biosynthesis,
- in production and maintenance of intracellular pressure,
- in temperature regulation by means of transpiration.

Water is mainly taken up by the roots in an osmotic process based on the difference between intracellular and soil solution pressure. The suction power of roots depends on the concentration of the cell sap, whilst the uptake of water also depends on suction pressure and salt concentration in soil water. Clay soils have a higher percentage of fine pores and therefore a higher suction pressure than sandy soils.

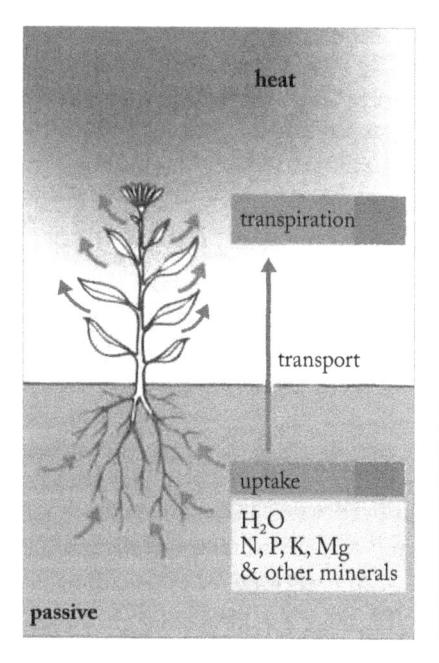

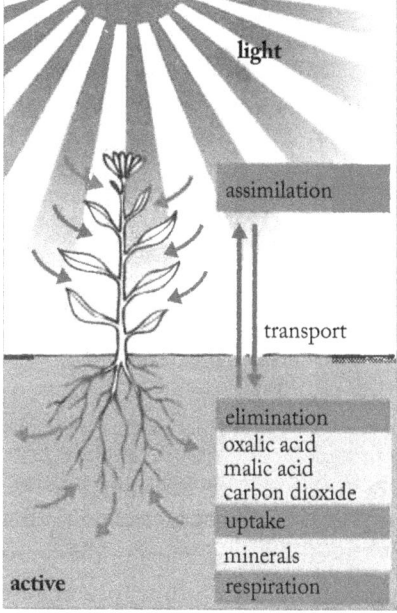

Figure 3. Active and passive mineral uptake.

The suction pressure of roots increases with the concentration of the cell sap, which the plant achieves by
- taking up minerals from the soil solution,
- supplying organic compounds and ions from its aerial parts
- controlled evaporation of water through stomata.

Other forces clearly also play a role in water uptake. The osmotic pressure gradient is limited to the cortex of the root, yet the water is still under pressure when it enters the vessels of the central vascular cylinder. Schumacher (1958) assumes that forces are involved that actively transport the water into the shoot. The energy required to produce the necessary root pressure derives from the plant's own metabolism. The energy required for transpiration, on the other hand, derives directly from solar energy with its effect on humidity and hence the water absorption capacity of the atmosphere. Water uptake is probably chiefly by active absorption at the root tip, with passive absorption due to transpiration taking second place (Hayward & Spurr 1943). When all available water in the immediate vicinity of the root has been taken up, plants follow the water by means of root growth.

Minerals are also largely taken up by the roots. Active uptake consists in the secretion of carbon dioxide, organic acids (e.g. malic acid) and other substances (complexing agents) to dissolve the required nutrients out of solid soil elements; these are then absorbed by ion exchange. Active mineral uptake is controlled by the plant's own metabolism, which explains why there is some degree of selection in mineral absorption into root cells. The plant secretions also create a favourable environment for specific microorganisms that help to dissolve the minerals.

Passive mineral uptake consists in dissolved salts being forcibly drawn into the root by transpiration pressure (e.g. nitrogen as NO_3). In this case the plant has little or no influence on the kind of minerals taken up. Compost rich in nutrients and well-rotted farmyard manure specifically encourage the growth and branching of roots, thus increasing the active root surface area. Their use enhances plant growth more than the concentration of soluble minerals would suggest.

At the end of his *The Power of Movement in Plants,* Charles Darwin wrote (page 573): 'It is hardly an exaggeration to say that the tip of the radicle thus endowed, and having the power of directing the movements of the adjoining parts, acts like the brain of one of the lower animals; the brain being seated within the anterior end of the body, receiving impressions from the sense-organs, and directing the several movements.' Phospholipids known as phosphatides found in some concentration in animal and human nerve tissues occur mainly at the root tip in plants. Their presence, the fine division of roots by branching, with rootlets penetrating the soil much as nerves do the human organism, and the way roots grow towards water and nutrients, all suggest that these substances have a function akin to sensory perception. The faculty of selective nutrient uptake and other features are clear indications that the root system plays the same role in the plant organism as the nervous system does in humans. Rudolf Steiner made frequent reference to this similarity.

water uptake	
passive	via transpiration based on physical laws
active	involving metabolic functions
nutrient uptake	
passive phase	as far as root cortex subject to physical laws
active phase	into inner part of root by means of ion exchange and metabolic functions
selective phase	plant roots select specific minerals, taking up greater relative proportion than present in soil water (legumes – relatively higher uptake of calcium, grasses – of potassium)
'chemical weathering'	release of protons, organic acids, complexing agents, etc., active lysis from mineral and organic part of crumb
Specific enzymes (permeases) permit uptake of organic molecules, e.g. plant protection agents such as antibiotics Release of organic substances to stimulate microbial activity in immediate vicinity of root 'The root system of a plant is in constant motion in its search for water (and nutrients).' (Larcher 1984)	

Table 1. Water and nutrients taken up by plant roots (Finck 1976)

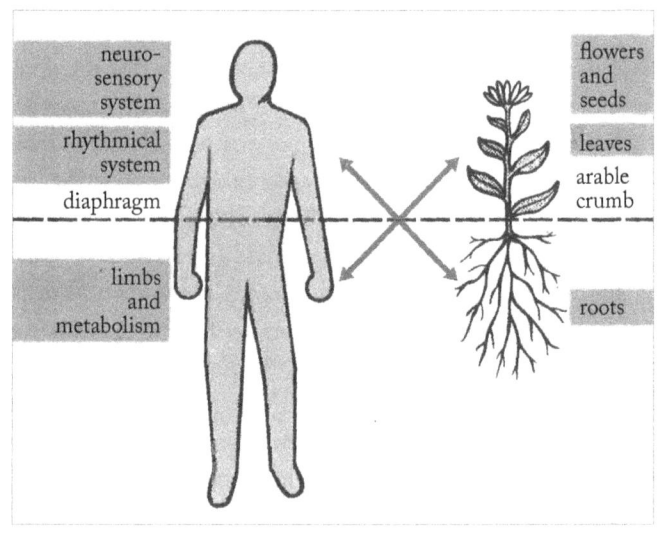

Figure 4. Three principal aspects of the human being and corresponding aspects of the plant.

> **Flowering region:**
> emissions (scent, pollen), movements of flowers (opening and closing, turning to light), discharge of pollen, fruiting process, corresponds to
> *Human limbs and metabolism.*
>
> **Leaf region:**
> 'inhalation' of carbon dioxide (oxygen and water), 'exhalation' of oxygen and water (carbon dioxide), rhythmical sequence of leaves, corresponds to
> *Human rhythmical system.*
>
> **Root region:**
> selectivity in uptake of nutrients, growth towards nutrients and water, presence of phospholipids, fine rootlets ramifying through soil, corresponds to
> *Human neurosensory system.*

Table 2. Flowering, foliage and root regions of the plant and their relationship to the human being.

3. Land Use

Woodlands are the natural form of vegetation in the Central European climate. Other forms of vegetation would only occur naturally in wet moors where nutrients are reduced, in salt marshes, on sand dunes kept moving by the wind, on rocks and in mountain areas. After the last ice age, the different tree species rapidly colonized areas once the ice had gone. When humans began to settle, woodlands gradually had to give way to farm land. Grazing animals and clearance by burning reduced woodland areas in a process that has continued into the present age, with glassworks, salt works and mines using large quantities of wood.

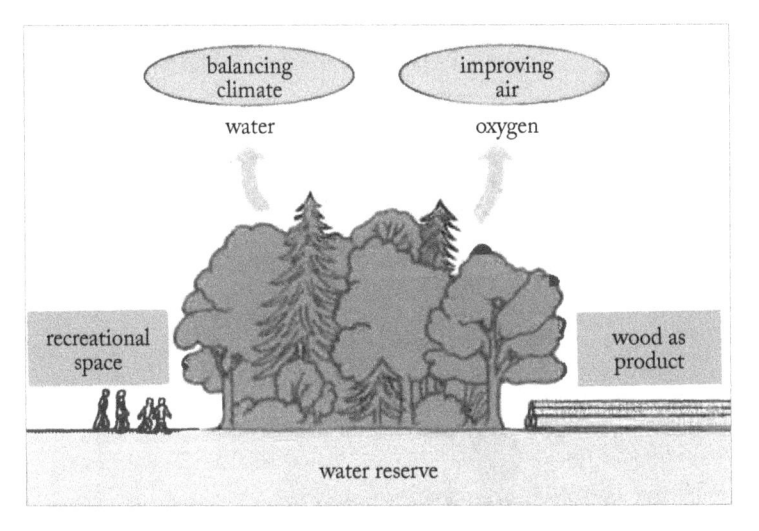

Figure 5. Functions and effects of healthy woodland area.

Today, woodlands are essentially only found in areas where soil conditions – stony or shallow soil, sloping ground or other problems – make agricultural use impossible. They do however continue to play a vital role as suppliers of wood and water reserves, balancing the climate and improving the air (amenity work). They add feature and distinction to the landscape and provide an area for healthy recreation. In recent times, woodland has been under serious threat from pollution, a side effect of advanced industrial development.

Agricultural land use is dependent on local conditions:
- soil type (sand, loam, clay, lime, moor)
- soil qualities (shallow or deep, heating up easily or only slowly, water holding capacity, drainage)
- climatic conditions (mean annual temperature and temperature profile, amount and distribution of precipitation)
- moisture (ground water level, surface water, lack of drainage)
- terrain (level, sloping, hilly).

This also determines whether an area is used for grassland, pasture or arable land, horticulture, fruit or wine growing.

4. Grassland

Grassland, also known as permanent and farmed grassland, means all areas in long-term use as meadows or pastures.
The most common reasons for putting down grassland are:
- high ground water level,
- risk of periodic flooding,
- high rainfall (900 mm or more per annum in Central Europe),
- areas where mists are common, with high humidity,
- heavy soil with high moisture content,
- sloping ground with shallow soil,
- areas distant from farm.

Grassland areas resistant to treading are used as pasture for dairy cows if near the homestead and for heifers if further away. Rotation of hay and grazing is the most intensive form of management. All other grassland areas are hayed once or twice a year depending on the site.

Moisture conditions in grassland need careful management. Water logging and long-term flooding must be prevented or dealt with, as they reduce fodder quality. The application of well-matured farmyard manure, careful husbandry and planned use result in high quality fodder with a richness of compostion that cannot be produced on arable soil.

There is a tendency, especially on arable farms, to pay little attention to such grassland as is available, thus neglecting a valuable source of fodder.

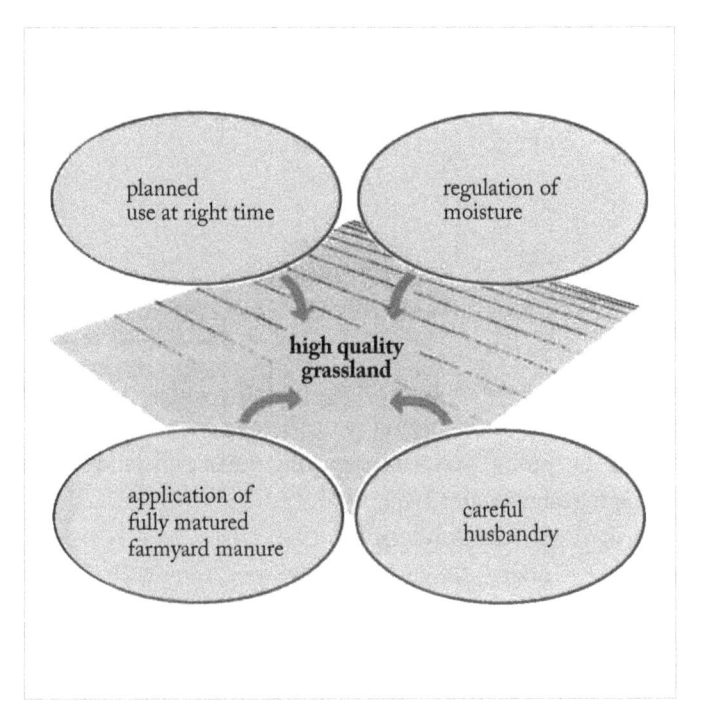

Figure 6. Preconditions for high quality grassland.

Cultivations

All grassland, whichever way it is used, needs the following cultivations to achieve a good dense turf with a satisfactory mixture of plants.

Harrowing should be reserved for exceptional circumstances. It is well known that damage to the turf does more harm than good. Rolling on the other hand is important for most grasslands, as it firms soil broken up by frost or other agencies. Pressure sensitive weeds such as cow parsley and hogweed are suppressed, coarser stalky grasses, other members of the carrot family, nettles, etc. inhibited, and tillering of close-growing grasses is encouraged. The roll must exert a pressure of at least 1 tonne/m^2 (1.5 psi) to firm the soil adequately. The time for rolling depends on moisture conditions. For good results, the soil should not be too wet nor too dry. If a heel is pushed firmly in the ground and water collects, the soil is too wet. It is too dry if the roll leaves no mark on the grassland or, even worse, dust is raised. Rolling generally encourages growth and a good yield in loose-structured grassland rich in humus. It does not normally have a positive effect on firm-structured grassland with low humus levels.

Molehills and ant hills are best levelled with a planker. Soil included with the fodder may cause digestive disorders in animals and have a negative effect on the fermentation of silage.

Regular clearing of drainage ditches is essential. The material removed from ditches makes a valuable addition when setting up farmyard manure for composting that is later used to fertilize pastures.

Potato haulms becoming available after harvesting may be used as a mulch on meadows and pastures. They protect the turf against drying out and frost damage and also have some value as fertilizer. In difficult sites and under unfavourable weather conditions such as drought and frost, a thin cover of straw or a very fine layer of fresh farmyard manure can also be helpful, especially on new sowings.

Manuring

Meadows and pastures are manured with well-rotted or almost humic compost or composted manure. Relatively small quantities applied annually are better than larger amounts used at longer intervals. Animals leave their excrement on pastures, so that 8–10 t/ha (3½–4½ tons/acre) a year is sufficient. Meadows used for hay and silage require 10–15 t/ha (4½–6½ tons/acre) annually.

The best time for compost applications is after the second or third grazing on pastures and after the first or an early second cut on meadows. The compost should be applied quickly and evenly, using a pasture harrow if necessary to rub it in. It protects the turf from drying out and stimulates root growth and rapid re-greening. Soil organisms are particularly active during the warmer part of the year and soon convert the compost into unstable and stable humus.

Grassland also responds well to an annual application of 15–20 m^3/ha (1600–2100 gal/acre) of well-fermented liquid manure (or dung liquor), either during the period of active growth when the soil is moist or towards the end of winter on top of snow cover, *provided the soil is frost-free.*

Liquid manure, whether drift or floating slurry manure, is also best applied at these times. Relatively frequent applications of small amounts of diluted liquid manure give better results than a single application of a large amount of undiluted, concentrated liquid manure. Annual amounts are 25–55 m^3/ha (2700–6000 gal/acre) of grassland, depending on the amount of water added. The addition of water prevents nitrogen loss due to evaporation. Three to seven parts of water added increase nitrogen utilization to between three and five times the value with undiluted drift manure and about twice the value with floating slurry manure (Schechtner 1969).

In early spring and late autumn, i.e. as close as possible to the winter season, horn manure is sprayed on frost-free grassland to provide 'vitalizing fertilizing power' (Steiner 1924). Spraying this preparation immediatly after utilization will encourage rapid greening. On pastures this is best combined with the operation to distribute the cow-pats.

When the whole area has a fairly even cover of green leaves, i.e. growth has reached 5–10 cm (2–4 in), horn silica is sprayed in good growing weather; this is repeated each time there is new growth. If the operation is omitted, reduced yields are to be expected. The area is not grazed cleanly.

> Small and frequent doses of well-rotted liquid manure favour a harmonious meadow flora. Heavy dressing with raw liquid manure leads to fouling with docks and weeds of the carrot family.

4. GRASSLAND

Meadows

In the western part of Germany, about 40% of agricultural land is permanent grassland, with 56% of this meadows, most of which are cut twice a year (figures for 1975).

It is possible to increase total yield and improve the composition of plant species by changing from two cuts to more frequent cuts and making the first cut earlier.

Meadows given regular applications of compost or dung liquor during the period of active growth generally show a richer and more balanced mixture of plant species and will often allow an extra cut. Compared to this the use of fairly large amounts of not fully matured farmyard manure and mineral nitrogen leads to imbalance in composition. The best time for the first cut is when there is a good balance between quantity of yield and quality in terms of crude protein and crude fibre content and digestibility; it depends on the stage of development reached by the plants.

The time for efficient utilization is a few days longer for a meadow with a good variety of species than for one with fewer species in which cocks-foot, meadow foxtail and false oat-grass predominate. Optimum utilization is limited to a few days for the first cut; going beyond this point means a bigger yield but rapid decrease in quality, particularly digestibility, and hence deterioration of fodder utilization.

Growing conditions are different towards the end of summer and in autumn, so that delay in cutting will not seriously reduce fodder quality, though it also does not increase yield.

On meadows resistant to treading, increased numbers of weeds can be effectively controlled by grazing once a year, if necessary with a cleansing cut to follow. Utilization is earlier in this way, so that

- seeding weeds (cow parsley, hogweed, etc.) are prevented from producing seeds capable of germination,
- tall grasses (cocksfoot, false oatgrass, etc.) are kept within limits,
- light-seeking close-growing grasses and white clover are able to grow better and compete better and

✻ plants sensitive to treading (autumn crocus, hogweed, etc.) are damaged and may even die.

Geith found that in a meadow grazed in three consecutive years, useful grasses and clover species increased from 32.5% to 80%, with weeds, poor quality grasses and bare areas reduced from 67.5% to 20%. The fodder quality was enhanced (Kiel 1954). The application of compost immediately after the grazing operation makes the result longer lasting.

Grazing meadows even if they are only resistant to treading for limited periods is therefore a good method of getting long-term improvement in grassland crops, with improved fodder yields and quality.

Pasture and pastures mown at intervals

Generally speaking, grazing is the more intensive form of grassland utilization. The most important point in management is to ensure that areas are grazed quickly and evenly, to allow for the necessary long rest periods in between. Grazing methods include continuous, paddock, rotational and strip (or rationed) grazing as well as rotation of hay and grazing.

Effective use of continuous grazing calls for detailed knowledge of potential performance on a given site (see Table 3 for approximate figures). This is the only way to get the right balance between the animals' fodder requirements and grassland regrowth potential. For grazing to be reasonably even it is important to put on different types of stock that complement each other, e.g. dairy cows, horses and breeding sows. There is still a risk, however, of the better, more palatable species being over-grazed, with the result that they are displaced by less valuable fodder plants and the value of the grassland crop goes down.

	paddocks	net yield	
	No.	kSE/ha	NEL/ha
continuous	1–2	400–2,000	4,000–20,000
paddock	4–8	1,400–2,800	14,000–28,000
rotational	9–18	2,000–5,000	20,000–50,000
controlled or strip*	movable fence	4,500–8,000	45,000–80,000

kSE kilo Starch Equivalent
NEL Net Energy Lactation
* daily fodder area per dairy cow *c.* 100 m² in spring; 30–50% more in autumn.
A dairy cow producing 4,500 kg of milk on average per annum will need *c.* 1,300 kSE (13,398 NEL) during *c.* 165 grazing days.

Table 3. Grazing methods.

For paddock grazing, the total pasture area of the farm is divided into four to eight paddocks. The animals remain in one paddock until it has been fully grazed, which is usually one or two weeks. They are then taken to the next paddock. The method is particularly useful for more distant pastures grazed by heifers, as it requires less labour. On the other hand fodder loss due to treading and lying down is relatively high, and a cleansing cut will be required after each grazing period.

Rotational grazing means a further intensification of utilization. The grassland is divided into between nine and eighteen paddocks, each of which can be grazed in three or four days and is then given a rest.

Controlled or strip grazing involves division into an even larger number of sections. The animals are moved daily or after some hours. This gives long rest periods and maximum fodder yields; grazing is to maximum level and the animals are offered fodder grown to a controlled stage. The method is however the most labour intensive. The animals are feeding on young pasturage low in crude fibre, which means that their feed has to be balanced by offering hay made from more mature plants, straw (possibly including undersown clover grass), bran, kibbled oats or other material with a high crude fibre content, which is best done at milking time.

Twice-daily moves to new grazing areas are the ideal for achieving short grazing periods followed by long rests. Manure left by the animals should be evenly distributed as soon as possible after grazing, spraying the horn manure preparation at the same time. If horn silica is then applied six to twelve days before the animals are put on again, new fodder will grow quickly and the animals will find it palatable and graze cleanly.

With feeding on young pasturage, supplementary high-fibre fodder is required.

Humified or well-rotted compost applied between grazing periods encourages the growth of valuable fodder plants.

All grassland areas resistant to treading and close to the homestead are best utilized by rotation of hay and grazing. Careful planning of utilization times – alternating early and late utilization, with reference to the growth stages of fodder plants – is the most effective form of management, since many meadow weeds are intolerant of treading and hard grazing. Conversely, plants that the animals find unpalatable are less likely to spread if cut at intervals.

Rotation of hay and grazing permits adaptation to extreme weather conditions in so far as areas designated for grazing can be left to mature for haying if they are too wet, with the sward sensitive to treading. On the other hand, an area designated for haying may be used for grazing if dry weather conditions have resulted in poor growth.

Another way of controlling and improving herbage composition on grassland is to use carefully planned rotation of hay and grazing over a number of years, rather like a good cropping sequence on arable land, combined with alternating applications of more and less rotted down farmyard manure, as shown in Table 4. Fresh dung liquor and farmyard manure will give lush growth of grasses and some herbs, e.g. caraway; well-fermented dung liquor and humified farmyard manure will favour the growth of leguminous plants and finer herbs such as common birdsfoot trefoil, slender trefoil, bush vetch, meadow vetchling, salad burnet, ribwort plantain and lady's mantle.

Year 1							
Manuring			humified compost		dung liquor		
Use	graze v. early	cut		graze		cut	graze
Year 2							
Manuring			dung liquor rotted-down compost				
Use	cut rel. late	graze		cut		graze	graze
Year 3							
Manuring			dung liquor		humified compost		
Use	graze moderately early	cut		graze		cut	graze
Manuring will not be necessary in year 4.							

Table 4. Management of hay and grazing rotation.

In 1952, I was able to see remarkable results from a change in grassland land management at the Talhof Farm.

Available pasture consisted of 10 ha (25 acres) of very shallow weathered Upper Jurassic soil on sloping ground. The dairy herd (19 cows on average) was put on for short-day grazing (*c.* three hours daily) from May 10 to October 1. During the summer, ten young animals and heifers were put on daily, usually after the cows. The cows were given sufficient home-grown fodder and meadow grass to satisfy before going on to the pasture in the mornings and again in the cowbarn in the afternoon. Pasture provided merely a good third of their total rations. The young animals also needed a small amount of additional feed.

Apart from some good grasses – mainly yellow oatgrass, smooth meadow grass, rough meadow grass and red fescue – the plant population consisted of less valuable species such as sweet vernal grass, crested dogstail, soft brome, some sedges and rushes. Remarkably, legumes and herbs were absent, except for daisies, ox-eye daisies, hoary plantain, wild carrot and a few other members of the carrot family.

One third of the pasture land was ploughed, used as arable land for two to four years and reseeded. Three years after reseeding we had the dread 'hungry years'. By then we had however built up a good store of compost made from farmyard manure, dung liquor, aquatic plants from clearing the banks of a nearby river, beech, hornbeam, oak and ash leaves from our mixed woodland area, plus a small proportion (10%) of sawdust. As far as possible – the sloping pasture land frequently cannot take vehicular traffic in wet years – all pastures were treated with 25 t/ha (11 tons/acre) of compost twice a year – in early spring and then in July and August. Horn manure and horn silica were used intensively, with five or six spraying operations during the period of active growth.

	1952	1957
grazing area	10 ha	11 ha
stock	30.6 CU	45.3 CU
grazing method	short-day 3.5 hours daily	strip day and night moved twice daily
days grazed	145	190
forages as % of total	c. 40	c. 85
feed requirement feed days/CU/ha – with supplementary feeds	444	782
– corrected for	177	665
supplementary feeds		
See text for explanation		

Table 5. Change to new method of grassland management at Talhof Farm.

After a transition period, cows and young animals stayed in the pastures day and night. Electric fencing was used to give the dairy cows a new feeding area twice daily. Within five years, an average of 26 dairy cows and 18 young animals were getting almost their whole food requirement from 11 ha (27 acres) of pasture land, with the grazing period extended to run from April 25 to October 31. Only 10–15% of extra feed (on a dry matter basis) was required.

To our surprise and delight the herbage composition also changed. Large numbers of leguminous plants, valuable grasses and herbs made their appearance – perennial rye-grass, meadow fescue, cocksfoot, some timothy, lady's mantle, caraway, burnet saxifrage, plenty of white clover, common birdsfoot trefoil, a perennial red clover similar to the Swiss alpine meadow type, black medick, bush vetch, meadow vetchling, occasional patches of lucerne or sainfoin, and on dry slopes also occasional kidney vetch.

When the grassland experts from the regional authority (Baden-Wuerttemberg) came on a tour of inspection, their unanimous opinion was: 'Having seen this, we believe you that your cows do not need homegrown high energy food, seeing the kind of pasturage you have.'

The pastures now have 10 t/ha (4½ tons/acre) of composted manure three times in four years, and 25 m^3 (900 cu ft) of dung liquor three times in four years. Paddocks that can take vehicular traffic are used for rotation of hay and grazing. Apart from grazing, the 100 ha (250 acres) provide 20–30 tonnes of first-class hay for calf and young bull rearing.

Fifteen years after reseeding, soil analysis gave the following results for 3.8 ha (9.4 acres) ploughed and reseeded:
- humus 6.8%, pH 6.9
- P2O5 38 mg/100 g
- K2O 57 mg/100 g
- Mg 7 mg/100 g
- total N 0.47%.

And for 3.9 ha not ploughed, compost applied:
- humus 7.4%
- pH 6.7
- P2O5 12 mg/100 g
- K2O 30 mg/100 g
- Mg 6 mg/100 g
- total N 0.52%.

In either case, no additional organic fertilizer, rock dust, calcified seaweed or raw phosphate had been added to the compost or the dung liquor.

Well managed grassland will therefore yield fodder that offers rich variety, balanced nutrition and excellent value. At the same time the animals are able to move freely in light and air. Health and a good and long life quality are the result.

Plants and animals work together to provide stable, permanent fertility in grassland soils.

Spring and autumn grazing technique

In Germany, active growth, until then slow and hesitant, becomes rapid at an almost explosive pace in the second half of May or in June. To utilize the best harvesting time, it would then be necessary to graze or cut the whole area within just a few days.

The problem can be overcome to some extent by starting to graze very early. Cows are put to pasture by the end of April or in early May, when the plants have reached a height of 5–8 cm (2–3 in); they are kept on full winter rations, with grazing limited to a few hours to begin with, increasing the time day by day. During the first five days, 30 cattle units will need about one hectare of grassland. When rapid growth has started, the area required per day will decrease rapidly, despite the fact that indoor feeding is progressively reduced. The method ensures that all paddocks designated for grazing will have been grazed once before the peak harvest time arrives.

The best time for cutting crops for silage is about ten to fourteen days before the hay harvest. If grassland use is divided to good effect into silage, hay and grazing, it will be possible to harvest most of the crop at peak time without getting major peaks in labour requirements.

Careful grassland management also means starting grazing in a different paddock each year; otherwise early pasturing will harm the plants and result in undesirable changes in plant population as well as reduced yields (see Table 4).

In autumn care must be taken that grazing areas are not cropped too short before winter comes, as this may lead to increased frost damage. Paddocks that are to be grazed first in spring are the first not to be grazed in autumn, so that the grass can make some extra growth. The height should not exceed 10 cm (4 in), however, as extended periods under snow or flood water may result in gaps due to rot. The weeds growing in such gaps will then reduce yields. Unless the ground is too wet, it is a good idea to distribute the excrement evenly once more after the final grazing period. This operation is best not done when there is danger of night frosts in spring.

Watering

Grazing animals need free access to fresh, clean water.

Automatic drinking fountains may be put on embankments, by drainage holes and in similar places. If there are no open water places, or if the water is too polluted, pipes may be laid to automatic drinking troughs in individual paddocks. Plastic pipes are less expensive and require less labour than metal ones.

If none of the above are available, a water cart serves the purpose. A new type of bowl has been developed for cattle; it meets the need for large quantities of water taken relatively fast (developed in Weihenstephan, manufactured by Gloeggler).

Black fibreglass barrels have proved useful, as algal growth is kept to a minimum. With these it is usually enough to give a thorough clean just once, at the end of the grazing period. In paddocks with electric fencing, an electric wire run round the barrel will prevent the animals from pushing or rubbing against the barrel; elsewhere robust protection will have to be fitted.

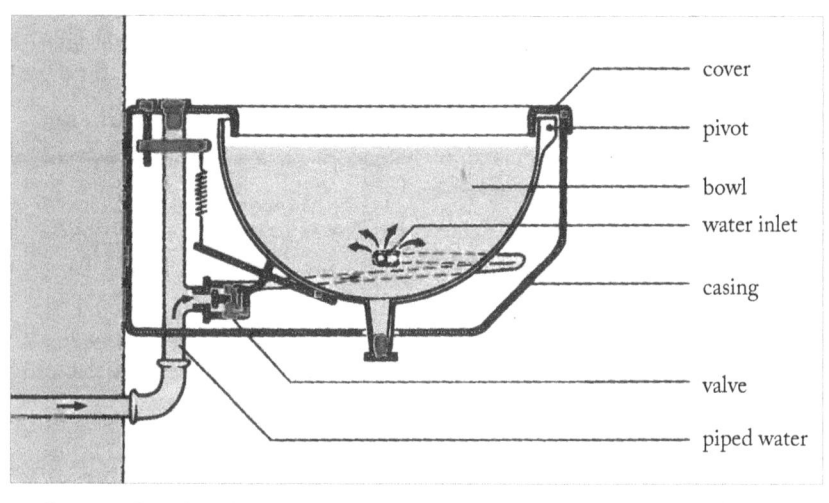

Figure 7. Drinking bowl for species-adapted watering of cattle (Gloeggler).

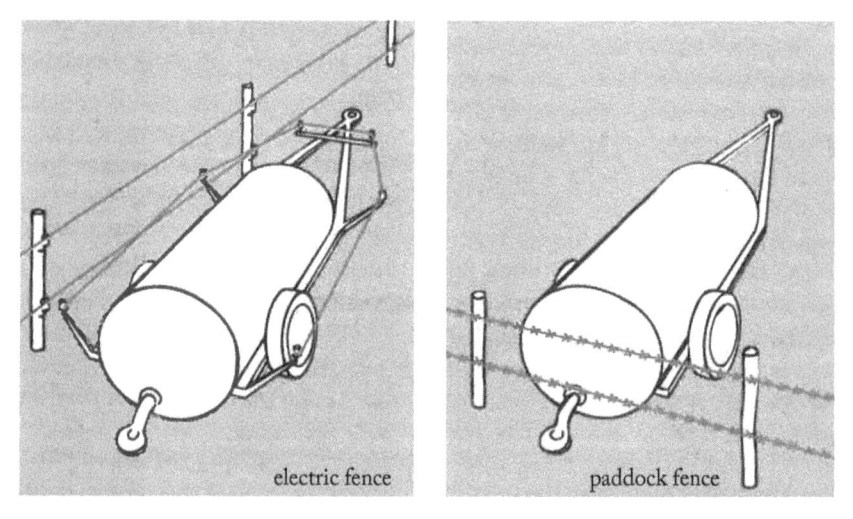

Figure 8. Protecting water barrel from damage by cattle.

Cleansing cut

If weather conditions are good and grazing is effectively managed there is usually no need to mow paddocks after grazing. If long wet periods make it impossible for silage or hay, the crop may go beyond maturity. In that case, grazing must be followed by a cleansing cut. Rainy periods are often followed by hot ones, so that the material, cut without a swath board, may be left on the ground to protect the sward.

Some plants will produce seed during long dry periods, which may mean an undue increase at the cost of other species. In this case, mow the paddock after grazing (with swath board), swath and collect the plants when wet with dew, before the seeds drop out. This applies particularly to caraway, which is further encouraged by farmyard manure. Cows that eat too many caraway seeds run the risk of abortion. It is therefore important to take extra care in composting and later using these residues, to make sure no seeds capable of germination get on to meadows and pastures and cause new weeds to grow.

Grassland weed control

Grassland plants are rated as weeds if they damage or seriously affect grazing stock, the turf or economic viability either singly or in massive numbers.

Management faults or omissions and site problems are the main reasons why undesirable plants appear; indicator plants may provide useful information on site conditions. Sour grasses, for example, indicate wet soil with poor drainage, kidney vetch and hoary plantain dry sites rich in lime, broom and ling a low pH, and quaking grass and ox-eye daisy soil low in nutrients (Table 6).

Weed problem due to	Control
Management errors	
cut too late	cut meadows fouled with weeds before seed-producing weeds come into flower; graze meadows resistant to treading once a year
frequent use with inadequate rest periods, grazing not properly controlled	regular use, with short grazing and long rest periods
injury to sward through wrong use of grassland harrow and putting heavy vehicles on wet ground	avoid injury to sward (no or only limited use of harrow, avoid putting vehicles on wet ground)
Site-related problems	
soil poor in nutrients low pH poor drainage periodic flooding frost breaking of mineral soil and looseness of humus rich soil	manure calcified seaweed drainage and keeping drainage channels clear rolling

Table 6. Causes of weed problems in grassland and control methods.

Some of the most important harmful plants and methods of controlling them are given below (see also Table 7).

A serious view must be taken of all poisonous plant species in meadows and pastures, as they are an immediate danger to grazing stock.

Marsh horsetail is poisonous when green, and in hay and in silage. It causes a marked reduction in milk production and loss of weight. The weed is controlled by draining the land, firming the sward by frequent rolling and intensive grazing followed by an aftermath, with the plant material removed.

Autumn crocus may cause fatal poisoning whether green, included in silage or dried. It does not tolerate treading and is therefore not found on grazed land. On grassland resistant to treading it will definitely disappear after a few years of intensive grazing. On grassland used for hay only, it helps to pull out the whole bunch of leaves in May (this removes the seeds that are just beginning to form

4. GRASSLAND

definite weeds	'relative' weeds
poisonous:	*massive populations:*
marsh horsetail autumn crocus marsh marigold creeping buttercup meadow buttercup	carrot family hogweed, cow parsley, chervil, ground elder, wild parsnip, wild carrot, angelica etc.
avoided by animals:	*plants difficult to utilize:*
creeping thistle woolly thistle	plants with ground rosette (dandelion, daisy)
stemless thistle stinging nettle ling tussock grass sedges (also mat grass) broad-leaved dock curled dock	*causing major losses:* bistort

Table 7. *Grassland weeds.*

with the leaves) and by repeated mowing when the plant flowers in autumn.

The poisonous species of the buttercup family – marsh marigold or kingcup, creeping and meadow buttercup – lose their toxicity on drying. Effective control is usually achieved by draining affected areas, cutting early for hay and then grazing repeatedly.

Massive occurrence of members of the carrot family (hogweed, cow parsley, chervil, ground elder, wild parsnip, wild carrot, angelica) usually relates to high level use of fresh farmyard manure and especially dung liquor or slurries. Numbers may be reduced as follows:

⚘ application of composted manure to which calcified seaweed or magnesian limestone and superphosphate has been added, 15 kg/ha (13 lb/acre) in either case, to balance the high nitrogen and potassium levels in the soil

- mowing before weed seeds are ripe
- if possible, grazing repeatedly.

Curled and broad-leaved docks are much more difficult to deal with. A relatively sure and probably also the only measure where growth is dense is to pull the plants out of the damp soil around St John's tide, i.e. between June 15 and 30.

If the above measures do not help, or if these or other grassland weeds are too numerous to pull out, it will be necessary to plough, providing soil conditions permit this.

Creeping thistle and stinging nettle are controlled by keeping growth down on pastures. The plants will gradually disappear as they cannot develop sufficient foliage for assimilation. Cutting down, using a scythe on small patches, may have to be done repeatedly for some time, sometimes a number of years.

The biennial woolly thistle and stemless thistle are easily got rid of by manuring the affected areas well and cutting repeatedly.

Ploughing and reseeding of permanent grassland

Before it is decided to improve a grassland area by ploughing it up, all other methods – different ways of manuring, different use or soil improvement – should be tried. It is vital to review the water situation. On dry or consolidated soil, the application of about 60–100 tonnes of compost per hectare (20–35 tons/acre) for two consecutive years is more likely to give good results than ploughing. Four or five applications of horn manure made during the period of active growth will markedly improve the growth of desirable fodder plants.

Year 1	grassland ploughed early July cultivation for two or three weeks oats and fodder vetch cut for fodder in autumn careful winter furrow
Year 2	early potatoes soil cultivation winter rye or wheat
Year 3	harvest summer furrow intensive harrowing and grubbing plenty of composted manure winter rape or Perko
Year 4	cut for fodder in spring (April/May) prepare seed bed for reseeding grassland reseed grassland grassland seed only or nurse crop, e.g. mixture of oats, peas and vetches cut nurse crop early in dry weather

Table 8. Example of cropping sequence after ploughing up grassland under conditions pertaining at Talhof Farm.

Ploughing is only indicated if the land is seriously fouled with perennial weeds or to cultivate uneven waste ground. Harmful weeds will only be completely removed by intermediate arable use. This does however mean considerable humus breakdown, which should be kept to a minimum and utilized to best effect by choosing the right cropping sequence. It is advisable to plough after a late first cut for hay, as reserves stored in weed roots will be low at that time. Success will depend on careful ploughing followed by thorough use of a heavy harrow and cultivator harrow. It is sometimes possible to get at least part of the roots to the surface, collect them and compost them in a separate heap. The compost should only be used once all root elements have humified completely.

Alternating thorough cultivation and fast-growing crops that give good ground cover needs repeating until there is no further risk of viable root elements causing new weed growth.

Seed mixtures for long-term grassland have to meet different criteria than the more short-lived clover and grass mixtures used for leys. The key criteria in devising a mixture are as follows:
- accurate knowledge and observation of site conditions,
- intended use (meadow, rotation of hay and grazing, intensive or extensive grazing),
- growth characteristics and competitiveness of species and varieties.

The herbage composition of grassland areas managed in the same way in the neighbourhood can provide helpful information on the choice of fodder plants for reseeding. Professor Klapp and others have done extensive trials to establish the varieties that are worth sowing and will 'take', and to assess changes in herbage composition over the years. Their recommendations for seed mixtures for reseeding permanent grassland have been published.

Seed quality must be first class, with preference given to locally grown product where possible. If purity and germinating capacity are low, the quantity needs to be increased accordingly, which also applies to unfavourable soil and site conditions. A fine, compact seed bed is important if the seed is to 'take' well. It has proved more effective to give the preceding crops a good application of compost, if necessary with the addition of calcified seaweed and possibly also raw phosphate, than to apply compost when reseeding.

Just as with rich clover mixtures on arable land, the method is to sow the larger seeds with the nurse crop, putting them in more deeply. The smaller round seeds are then drilled travelling at an angle of 30° to the previous direction, after which the roll is used. This will rapidly give a close sward.

Fine round seeds are more apt to separate out from mixtures. They should not be taken to the field in the drill, therefore, but added in smallish amounts at a time, giving the mixture a good stir at intervals. In areas where the spring seasons tend to be dry, drill very early and at least one centimetre deep, or drill without a nurse crop in July or August. A nurse crop that is not too dense protects the tender grass

4. GRASSLAND

	cool, fairly dry site		medium and variable moisture		abundant moisture		higher altitudes	
	meadow	pasture	meadow	pasture	meadow	pasture	meadow	pasture
false oatgrass	10						5	
cocksfoot	4	2	2				2	2
meadow fescue	7	7	20	14	14	18	14	14
timothy	2	2	2	2	2	3	4	4
smooth meadow grass	4	8	3	3	2	2	3	3
perennial ryegrass		3		10				
red fescue			3	3	2		4	4
meadow foxtail			3		6			
fiorin					1	2		
common birdsfoot trefoil	3	3		4			3	3
white clover	2	2	2			1		2
Alsike clover			3		2		3	
greater birdsfoot trefoil					2	1		

Perennial ryegrass and red clover when sowing as spacer, using 2–3 kg/ha; absolutely essential to keep new growth short, otherwise threatens to take over.
Yellow oatgrass: 1–2 kg/ha a valuable addition on relatively good, dry or medium moist mineral soil.
Canary grass: use 1–2 kg/ha on flood and irrigation meadows.
Lucerne (alfalfa): use 2–4 kg/ha on predominantly dry land suitable for lucerne (always of native origin!)

Table 9. Sample seed mixtures for permanent grassland (Klapp 1951). Amounts in kg/ha (multiply by 0.9 to convert to lb/acre)

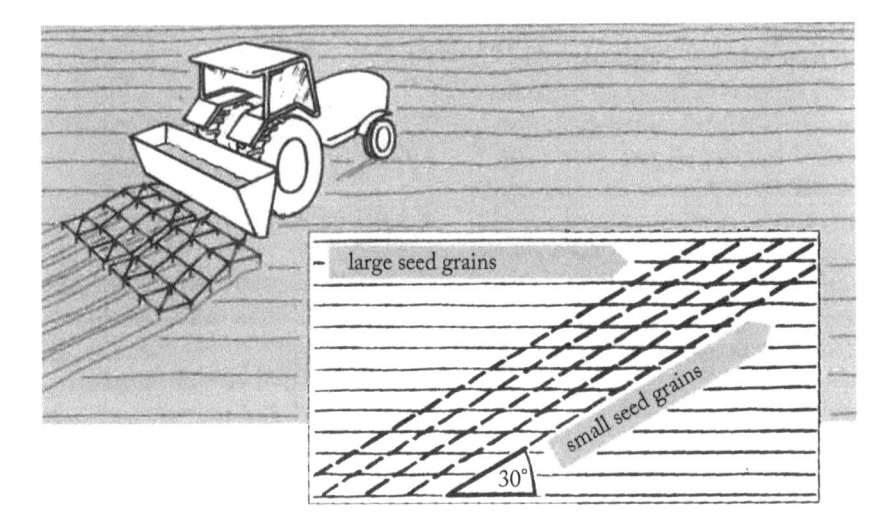

Figure 9. System for drilling seed of different size for seeding grassland and grass/clover mixtures.

and clover seedlings from wind, hoar frost, heavy rain, direct sunlight and competition from weeds. The nurse crop needs to be harvested when the ground is dry and also early enough, so that the seedlings will have all the light and air they need. A silage cut is best, as the fodder is taken away immediately, requiring no other field operations. If the soil can take vehicle pressure at this point, leaving hardly any tracks at all, a thin veil of very fine compost and at the same time an application of horn manure spray will make a tremendous difference in encouraging the young crop to grow on.

Hedges and trees in grassland

Hedges planted along drove roads and around paddocks provide protection from wind, rain and bad weather and above all also strong sunlight for the stock; they also provide food for variety and health; many hedgerow plants are rich in minerals, vitamins and trace elements;

- nesting places for insect-eating birds;
- a habitat for small animals such as shrews, hedgehogs, weasels, stoats and polecats that help to control insects and mice (field mice are natives of the steppes and do not go near hedges).

The investment needed to plant and maintain hedges does give maximum benefit. No good grassland farmer will spare the effort and expense for something that benefits farm animals, landscape and environment. When an agricultural area does not include woods or coppices, life and variety are added by planting parts of pasture or low-grade arable land with groups of trees and shrubs as well as hedges. A landscape cultivated in this way offers benefits and protection to stock, makes human beings feel more at home, and also enhances the recreation value for visitors to the countryside.

5. Crops and Cropping

A look at the history

Crop growing and plant breeding began when tribes started to settle. Advanced early civilizations in the Iranian uplands and Mesopotamia, along the Indus in Pakistan, the Hoangho in China, the Nile Delta and in Mexico brought intense development of farming methods and plant breeding.

Zarathustra, the great initiate of the ancient Persian civilization, instructed people in soil cultivation: 'Taking a golden dagger, he scratched the soil.' The heat of the sun and light-filled air penetrated the soil which had thus been torn open; arable soil was created, with the necessary conditions for breeding the cereal species and many other cultivated plants.

Archaeologists have found emmer (an early form of wheat) and barley in Arpachiyah (Northern Iraq, Assyria), emmer, barley and vetch (*Vicia* species) in Merimda Beni Salama, Ma'asari, Maadi and Faiyum (Egypt), all dating back to about 4000 BC; finds of barley, flax, lentils and a crucifer species (mustard or cabbage) in Sumer and in Uruk date back to the Sumerian civilization of 3100 BC. It is evident that plant breeding and a cereal diet were an important precondition for the development of those early civilizations. The earliest traces of cereals in Europe found near Lake Mälaren in Sweden date back to about 3600 BC.

Cultivated wheat, barley, millet, rice and maize appeared suddenly and always hand in hand with equally rapid progress in civilization. The many different breeds of domestic animals and useful plants (e.g. many varieties of fruit) that we have today largely go back to that early period of civilization in the fourth and fifth millenia BC.

Further new development has really only come in the last two centuries, i.e. about six thousand years later. The science of genetics and its practical application have led to the development of many new plant varieties and animal breeds. Scientific observation and the knowledge gained in physiology and biochemistry are becoming increasingly more refined and differentiated.

The introduction of modern agrochemical, farm and labour management methods has had progressively more negative results, especially in the last three decades:
- signs of rapid degeneration and increasing susceptibility to pests and diseases in plants,
- increasing loss of fertility and the appearance of new diseases in animals.

The use of new seed varieties almost every year, the increasing use of chemicals on plants, rapid rotational grazing and intensive use of prophylactic medication on animals provide only short-term solutions and do not deal with the causes.

There is increasing awareness that new ways will have to be found and followed, and this is a major challenge for the theory and practice of agriculture. First attempts made in plant breeding, for example, have shown that it will require enormous effort to find new methods that will be a real help in the foreseeable future. Existing plant species and varieties will need careful nurturing. When a particular variety has proved successful on a farm, it is advisable to follow the methods given in the section on seed production. Optimal cropping sequences and the decision to grow only crops that suit the habitat support our efforts to develop sound farming practices.

The sidereal system

The oldest known form of crop growing is the sidereal system (from the Latin *sidus,* star, 'determined by the stars').

Wiljams (1949) states: 'The Romans took the fully developed sidereal system from the Greeks, who had adopted it from the Egyptians; they, in turn, had taken over the complete system used by the peoples of the East. Anything before this is lost in the mists of history.'

'The system is to sow winter rye or mustard every one or two years after the crop has been harvested, ploughing in the rye after shooting and the mustard after flowering in late autumn. The Egyptians replaced mustard and rye with Egyptian clover *(Trifolium alexandrinum),* the people living in what today are the Central Asian Republics with mung bean *(Phaseolus mungo)* and those in Tajikistan with field pea (*Pisum sativum* ssp. *arvense*).' (Ruebensam & Rauhe 1964).

It may be assumed that the system was evolved by people with vast knowledge of the connection between stellar activities and everything that happens in the kingdoms of nature on earth.

This explains why the cultivated plants of today evolved so rapidly when farming began in 6000–4000 BC. Cultivated plants were bred from wild plants by taking account of specific planetary conjunctions and oppositions.

In biodynamic agriculture, astronomical researches are in progress to rediscover the skills and knowledge of the plant and animal breeders of ancient Persia and develop them further on the basis of modern science and the conscious mind of today.

In one particular experiment it proved possible to achieve genetic changes in cultivated plants by merely sowing and replanting them at the times of specific conjunctions and oppositions. A vast field has opened up for scientists with unbiased minds who have the courage to take this up.

Two and three-field system

It seems that the sidereal system was lost in Roman times; the ancient Germans either lost track of it or did not have it at all.

Tacitus wrote that, unlike the Romans, German tribes were using a strict rotation of crops. A two-field system of alternating cereal crops and fallow would be used in poor conditions, a three-field system of winter cereal, spring cereal and fallow in areas with better soil and climate.

Properly managed fallowing improves humus levels and stable crumb structure and keeps weeds under control. In the eighteenth and nineteenth centuries red clover and other leguminous plants were introduced, as well as potatoes, beet, oil plants, etc., and this gave an improved three-field system, with the new crops replacing the fallow. Increased fodder production meant larger quantities of farmyard manure, and this, together with raised nitrogen levels from the inclusion of leguminous crops, led to increased yields.

The improved three-field system also included roots and tubers: winter cereal (rye), spring cereal (oats), red clover, winter cereal (wheat), spring cereal (barley), root or tuber (potatoes).

> Carefully planned cropping sequences are a major factor in farm productivity.

Crop rotation

The improved three-field system marked the transition to crop rotation. Two-course rotation alternates non-cereals and cereals, double rotation has non-cereals twice and cereals twice. Yields are markedly better with the double system.

In polycrop rotation, cereals are grown after three non-cereals, e.g. two crops of grass and clover ley, potatoes or barley.

6. Cropping Sequence

The productivity of farms with a relatively high proportion of arable land depends very much on a properly planned and consistently used cropping sequence. It takes a long time to work out the ideal rotation for a given site, calling for careful observation and a continuous learning process.

When a sequence has proved right for a particular site, stability and level of yield and soil improvement will increase with the second and subsequent cropping sequences.

It is often necessary to use two, and on large farms even three or more cropping sequences, as individual plots

- differ greatly in soil quality,
- the nature of the ground (stony, sloping, wet) may present problems with the use of machines,
- growing conditions are not right for certain crops, e.g. vegetables, and
- it may be necessary to have intensive cropping sequences closer to, and less intensive ones further away from, the homestead.

Percentage areas

The percentage of cultivated or arable land determines utilization as regards the relative area given to individual crops in a particular year. The best possible cropping sequence, i.e. the sequence in time over a number of years in a particular field, has to be evolved with this in mind.

Points to be considered when establishing relative areas are:
- fertilizer and fodder requirements
- growing soil-improving crops
- management aspects and work programme
- weeds and pests
- compatibilities

Fertilizer and fodder requirements

The two are closely connected. The home-grown fodder requirement, particularly leguminous plants, is the basis for any cropping sequence; in conjunction with livestock numbers it also determines the volume of manure and dung liquor (fermented liquid manure). The combined fertilizing power of these must be such that soil fertility is maintained and if possible even enhanced under existing site conditions. Top class soils usually require only 0.4–0.8 CU/ha (0.15–0.3 CU/acre) of farmland. If more roots, tubers or vegetables are grown (more than 25% of arable area), about 1 CU/ha (0.4 CU/acre) will be required. The poorer the soil and the climate, the greater will be the amount of manure required and therefore also the area needed for fodder. Generally 1–1.5 CU/ha (0.4–0.6 CU/acre) of farmland is generally adequate for efficient management.

Once the number and type of animals needed to produce the required amount of manure has been established, it is possible to calculate the amount of fodder required. Yields from absolute grassland and fodder grown as catch crops (where conditions permit) are calculated to work out how much fodder and root plants need to be field-grown. Catch crops grown for green manuring will help to meet manure requirements.

Soil-improving crops

A low percentage area of absolute grassland makes it easier to include soil-improving plants in the cropping sequence since many fodder plants, and particularly legumes, are in this category. It should be remembered that crops harvested green have a better preceding crop value, as the fine rootlets are lost with increasing maturity. Root and stubble material from plants harvested at maturity also has a lower C/N ratio, which makes the rotting down of these residues more difficult. Mustard and sunflowers grown as fodder thus rank as good preceding crops except when harvested ripe. Even with legumes such as peas, field beans, vetches and clovers, crops usually have a much lower preceding crop value if grown for seed or pulses than as green fodder and catch crops, the reason being that nitrogen from the root nodules is used to produce seed protein.

Economic and labour aspects

Crops to be considered for the remaining arable area are mainly such as thrive in the given conditions, are in demand and have a market that is within reach. This is where economic considerations are paramount. Preference is given to crops where the balance between expenditure and yield ensures long-term economic viability for a sound, productive farm unit.

Gross margins range from €2,500 to €9,000/ha (1989 values) for individual crops, with 20–250 hours labour/ha, with corresponding variation in utilization of labour hours (gross margin per working hour). It is evident, therefore, that the work programme needs to be planned. Available labour and the distribution and/or concentration of work in specific time periods will soon set a limit to the inclusion of crops that sell at a high price but also require much labour.

Seasonal variations in available days depend on the climatic conditions in different regions. The hours given in Table 10 represent mean values and major deviations are bound to occur.

Sales potential must also be taken into account. There is a definite limit to direct consumer sales; usually a fixed group of customers feels connected with and responsible for a particular farm, its crops, animals and people. 'Agricultural communities' connected with individual farms are a recent development.

Crop	yield	work required	gross margin	
	t/ha	hours/ha	€/ha	€/hour
Winter wheat	4.5	20–40	2,500–3,000	62–150
Potatoes	30.0	180–220	3,000–4,000	14–22
Carrots	45.0	160–250	7,500–9,000	30–57

Sale of winter wheat: 75% direct sale, 25% fodder;
Sale of potatoes: 60% direct sale, 14% seed, 26% fodder;
Sale of carrots: 10% direct sale, 90% juice.

Table 10. Gross margins for wheat, potatoes and carrots. Gross margin is at 1989 value, so merely for comparison.

Period	SC	R–H	CH	RH	LA	Total
days available	16–36	21–38	41–48	27–56		
winter wheat	5	4	12	10	4	35
potatoes (medium early)	32	8	1	90	72	203
carrots	4	30–90	25–45	42	60	161–241

SC = spring cultivations; R–H = cultivation of row crops/haying;
CH = cereal harvest; RH = row crop harvest;
LA = late autumn cultivations

Table 11. Level and distribution of labour required (in hours/ha)

Weed and pest control by use of cropping sequences

Weed and pest control is influenced to a variable degree by the position of crops within the cropping sequence. With monocultures, the presence of a number of weed species does not necessarily have a negative but may indeed have a positive effect on growth. This only applies, however, if the weeds remain small, do not act as vectors for harmful organisms, have different requirements where soil structure, nutrient supply and root space are concerned, and do not make cultivation and harvesting more difficult.

Crops with similar requirements that are grown in quick succession encourage massive populations of individual weeds or weed combinations. Serious fouling with persistent weeds calls for special cleansing sequences. Wild oats, for example, may become difficult to control if spring cereals predominate in cropping sequences. They will be markedly reduced if not eradicated by not growing spring cereals or winter wheat for three, or even better four years. Winter rye is the most powerful competitor and most able to displace weeds; it firmly keeps down wild oat.

Several years of lucerne (alfalfa) or clover and grass mixture will usually get rid of thistles.

Couch grass needs a lot of light and therefore does not tolerate two successive crops of an oat and legume mixture within a year, as these quickly provide dense ground cover.

Cereal crops grown in quick succession encourage take-all and similar fungus diseases. Pests also tend to increase (beet eelworm and clover stem and root eelworm) with successive crops of the same family, e.g. cabbage family (cabbage, swedes and rape) or pea family (white and red clover, crimson clover, lucerne, pea, vetch or field bean).

Nematodes may persist for a long time in the soil, but a well spaced cropping sequence is the best method of prevention. Diseases and pests are frequently due to intensive growing of crops not entirely suited to the site, usually for commercial reasons. The result is reduced resistance. Another way of putting it is to say that weeds and pests make their appearance when an imbalance has to be corrected in the biological partnership between soil and crop. Other plants and organisms – in this case weeds and pests – then find better growth conditions than the weakened crops. Resistance is increased and crops are better able to withstand poor weather conditions such as extremely dry conditions alternating with wet periods if crops well suited to the site are grown in well thought-out proportions, well-rotted manure treated with the biodynamic preparations is used that will not force growth, and the biodynamic sprays are also applied.

6. CROPPING SEQUENCE

Compatibilities

Some crops are self-compatible, others are self-*in*compatible. This latter group includes crops that will not thrive if grown again immediately or after only a short interval. Pests and weeds tend to increase, but even without this, yields will definitely be reduced. Problems of this kind are avoided by observing the required intervals.

Clover sickness is unlikely to develop if different legumes are grown in rotation.

self-compatible	self-incompatible	intervals (years)
rye	sunflowers	7–8
maize	flax	7–8
millet	red, Alsike, crimson clover	6–7
potatoes (not in nematode infested areas)	sainfoin	6–7
	sugar and fodder beet	5–6
white clover	cabbage varieties	5–6
hairy vetch	peas	5–6
fodder grasses	oats	4–5
yellow lupin	rape	4–5
soya bean	wheat	2–3
dwarf bean	Barley	2–3
serradella	potatoes (in nematode infested areas)	2–3
Examples of incompatible sequences		
barley – wheat oats – barley sugar beet – rape fodder beet – rape cabbage varieties – rape flax – peas	poppies – potatoes red clover – serradella red clover – lucerne (alfalfa) legumes – peas field beans – clovers	
Wheat after barley is incompatible, but barley after wheat is compatible. All other pairs are incompatible either way.		

Table 12. Examples of cropping sequence compatibilities.

When a cropping sequence involves frequent use of clover species and other legumes, clover sickness may develop. Stem rot, lesser broom-rape, stem and root eelworms and other problems arise. Seedlings are slow to develop and pea and bean weevils may destroy the whole crop. The plants wilt quickly in dry weather and are more susceptible to mildew and rust. Growth is stunted and winter hardiness reduced. The cropping sequence needs to be changed when such problems arise.

Catch cropping permits a more rapid succession of self-incompatible plants. 'The primary reason would seem to be that root and stubble residues from the catch crop have a higher nitrogen content and hence a lower C/N ratio than residues from the main crop and are therefore broken down more quickly by soil organisms.' (Ruebensam & Rauhe 1964). Conversion is faster in active humic arable soil than in unenlivened soil. Active soils have a better 'ability to digest' and there are fewer problems with cropping sequences.

Working with living things, we find over and over again that one cannot apply rigid principles in every situation, and this also holds true for cropping sequences. Intensive use of mineral fertilizers and synthetic plant protectives do make it possible to grow incompatible crops consecutively and to continue to do so for some time, but a reaction is bound to come sooner or later:

- expenditure will become too high in relation to product, or
- yields will be less reliable, with soil structure and fertility decreasing rapidly.

With the specific measures that are part of the biodynamic method, experience has shown that some of the principles relating to cropping sequences lose significance. Legumes, for example, will grow extremely well in soil that has had biodynamic care. If calcium and phosphorus are adequate, red clover and other legumes and clovers can be grown in quick succession and will give excellent yields for many years – more than 50 years on the Talhof Farm. Applications of humified farmyard manure allow peas to be grown as a vegetable twice in succession, with good results.

6. CROPPING SEQUENCE

Principles and terminology

Before giving examples of cropping sequences, some of the principles and terminology require discussion.

Terminology

M = main crop, C = catch crop, P = preceding crop, F = following crop

- *Main crop:* Occupies the field during main part of active growth period and provides main part of annual yield.
- Example: winter and spring cereals, sugar beet.
- *Second crop:* Main crop grown after an early cleaning crop.
- Example: early potatoes followed by marrow-stem kale, in favourable sites followed by beet seedlings.
- *Catch crop* Grown for fodder or as green manure between two main crops to prevent leaching of minerals.
- Example: winter barley (M) – Landsberg mixture (see page 190) (C) – maize for silage (M) – winter wheat (M) – spring rape, green (C) – oats
- *Preceding crop:* Immediately precedes another crop.
- *Following crop:* Immediately follows another crop.
- Example: peas (P) – winter wheat (F).

Every plant has slightly different requirements as regards soil, pH, water and nutrient supplies, i.e. growing conditions. On virgin soil, characteristic plant communities develop, with the species supporting, complementing or displacing and suppressing one another. Such a relationship in a given space will only prove successful in special cases such as leys, mixed spring crops and maslin, mainly if grown for fodder. On the other hand it is possible to let crops that complement or support one another follow each other in time. This is taken into account in cropping sequences, so that plants with diametrically opposed properties are grown in sequence:

- nitrogen-hungry plants and nitrogen providers,
- humus-hungry plants and humus providers,

- plants that reduce and those that help stable crumb,
- shallow-rooted and deep-rooted plants.

The examples given in Table 13 serve to demonstrate this, though it is not usually possible to take account of all four properties at once.

	nitrogen	humus	tilth	root space
year 1: ley	++	++	++	++
year 2: wheat	– –	–	– –	–
vetch and rye (catch crop)	++	+ –	++	+
year 3: potatoes	– –	–	++	+
year 4: peas	++	+ –	++	–
year 5: oats	–	–	– –	+
++ positive effect/deep-rooted + – neutral	– – depleting + moderately deep-rooted		– negative effect/shallow-rooted	

Table 13. *Complementary qualities in a cropping sequence.*

Preceding crop value and effect

A distinction should be made between:
- Preceding crop value: effect of a particular preceding crop on various crops that follow,
- Preceding crop effect: effect of different preceding crops on a particular crop that follows.

The requirements of crops and their effect on following crops determine their position in the sequence. Legumes, for instance, have a high preceding crop value even if grown as a catch crop because they fix nitrogen and are good producers of organic matter; potatoes do extremely well on this, for example. The preceding crop value of potatoes at the other hand is due to the intensive cultivation and harvesting operations that result in weed-free soil with crumby texture. It then usually needs only a culti-harrow or cultivator to prepare the seed bed for winter cereals.

6. CROPPING SEQUENCE

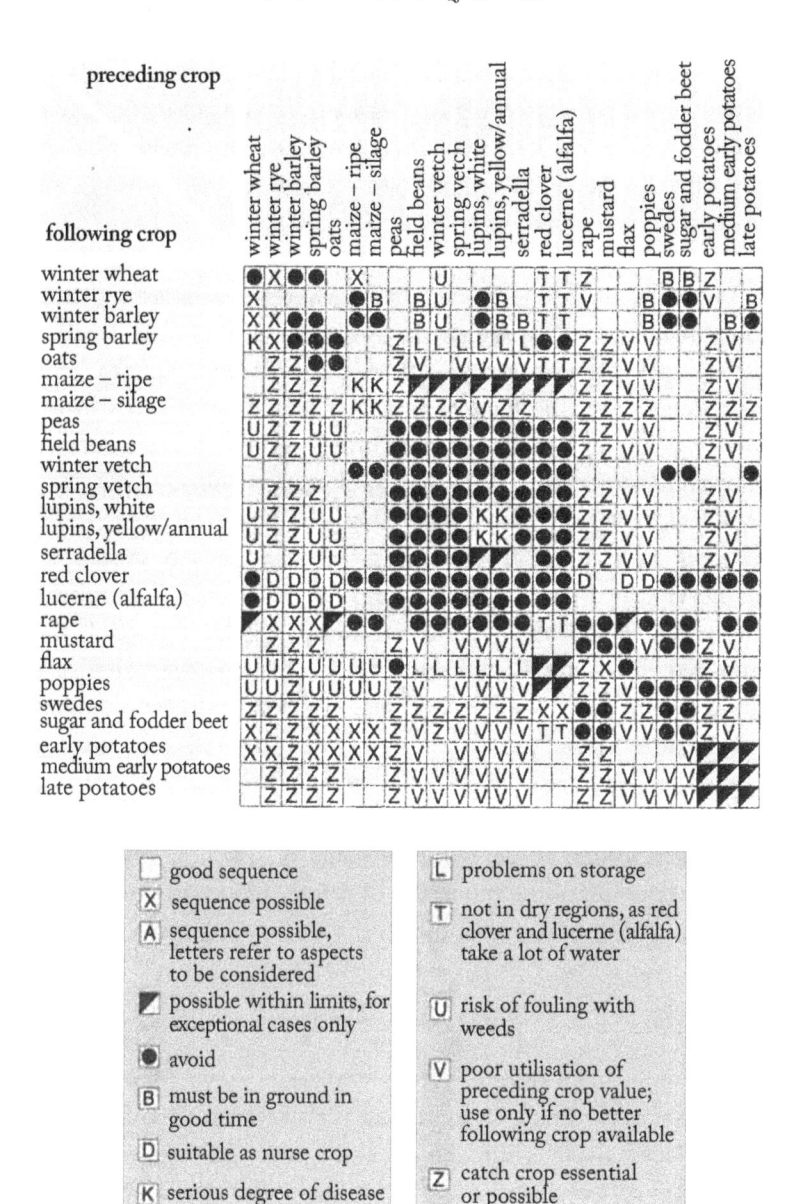

following crop \ preceding crop	winter wheat	winter rye	winter barley	spring barley	oats	maize – ripe	maize – silage	peas	field beans	winter vetch	spring vetch	lupins, white	lupins, yellow/annual	serradella	red clover	lucerne (alfalfa)	rape	mustard	flax	poppies	swedes	sugar and fodder beet	early potatoes	medium early potatoes	late potatoes
winter wheat	●	X	●	●		X			U						T	T	Z				B	B	Z		
winter rye	X		●	B		B	U		●	B					T	T	V				B	●	●	V	B
winter barley	X	X	●	●		●	●		B	U				●	B	B	T	T			B	●	●		B ●
spring barley	K	X	●	●		Z	L	L	L	L	L	L	L	L	●	●	Z	Z	V	V			Z	V	
oats		Z	Z	●	●		Z	V		V	V	V	V	V	T	T	Z	Z	V	V			Z	V	
maize – ripe		Z	Z	Z		K	K	Z	▨	▨	▨	▨	▨	▨	●	●	Z	Z	V	V			Z	V	
maize – silage	Z	Z	Z	Z	Z	K	K	Z	Z	Z	Z	V	Z	Z			Z	Z	Z	Z			Z	Z	Z
peas	U	Z	Z	U	U			●	●	●	●	●	●	●			Z	Z	V	V			Z	V	
field beans	U	Z	Z	U	U			●	●	●	●	●	●	●			Z	Z	V	V			Z	V	
winter vetch						●	●	●	●	●	●	●	●	●					●	●					●
spring vetch		Z	Z	Z				●	●	●	●	●	●	●			Z	Z	V	V			Z	V	
lupins, white	U	Z	Z	U	U			●	●	●	●	K	K	●	●	●	Z	Z	V	V			Z	V	
lupins, yellow/annual	U	Z	Z	U	U			●	●	●	●	K	K	●	●	●	Z	Z	V	V			Z	V	
serradella	U		Z	U	U			●	●	●	●	●	●	▨			Z	Z	V	V			Z	V	
red clover	●	D	D	D	D	●	●	●	●	●	●	●	●	●		D		D	D	●	●	●	●	●	
lucerne (alfalfa)	●	D	D	D	D																				
rape	X	X	X	X				●	●	●	●	●	●	●	T	T			●	●	●	●	●	●	
mustard		Z	Z	Z				Z	V		V	V	V	V					●	●		●	Z	V	
flax	U	U	Z	U	U	U	U	●	L	L	L	L	L	L	▨		Z	X	●				Z	V	
poppies	U	U	Z	U	U	U	U	Z	V		V	V	V	V			Z	V	●	●			Z	V	
swedes	Z	Z	Z	Z	Z			Z	Z	Z	Z	Z	Z	Z	X	X	●	●	Z	Z	●	●	Z	Z	
sugar and fodder beet	X	Z	Z	X	X	X	X	Z	V	Z	V	V	V	V	T	T	●	●	V	V	●	●	Z	V	
early potatoes	X	X	Z	X	X	X	X	Z	V		V	V	V	V			Z	Z				V		▨	
medium early potatoes		Z	Z	Z	Z			Z	V	V	V	V	V	V			Z	Z	V	V	V	V			▨
late potatoes		Z	Z	Z	Z			Z	V	V	V	V	V	V			Z	Z	V	V	V	V			▨

- ☐ good sequence
- X sequence possible
- A sequence possible, letters refer to aspects to be considered
- ▨ possible within limits, for exceptional cases only
- ● avoid
- B must be in ground in good time
- D suitable as nurse crop
- K serious degree of disease or pests possible
- L problems on storage
- T not in dry regions, as red clover and lucerne (alfalfa) take a lot of water
- U risk of fouling with weeds
- V poor utilisation of preceding crop value; use only if no better following crop available
- Z catch crop essential or possible

Figure 10. Preceding and following crops (Ruebensam & Rauhe 1964, based on Koennecke).

57

As one would expect, a lean and miserable crop with many weeds will always have a poor preceding crop effect as it does not produce good crumb structure, has a low root mass, provides poor ground cover and offers the likelihood of a serious weed problem for the following crop. Koennecke's preceding crop schedule (from Ruebensam & Rauhe 1964) helps to gain a clear picture.

Preceding crops have a good or bad influence on following crops through the condition which they leave in the soil. The schedule merely shows the immediate effect, and it is important to know and observe also the long-term effect, i.e. the extent to which a particular crop reduces humus and nitrogen levels in the soil. Potatoes grown in soil fertilized with farmyard manure, for example, are a good preceding crop for almost all other crops except poppies. In the long term, however, potatoes given an excessively great share in the sequence have a definite negative effect on soil fertility as they require large amounts of nitrogen. It is a general principle that crops requiring intensive cultivation have a negative preceding value because they deplete the soil of organic matter. If they dominate too much in a rotation, the balance must be redressed by growing sufficient other species with long-term soil-building properties, e.g. a ley of several years.

To help readers to develop some kind of system, a brief review is given opposite of the properties of cereals and grasses on the one hand and non-cereals on the other.

Cereals and grasses

The group includes all cereals and grass species grown for seed. They rank lower as preceding crops because they leave a poor soil structure and seriously deplete nitrogen stores if grown for high yield. The differences between species and varieties are relatively large.

6. CROPPING SEQUENCE

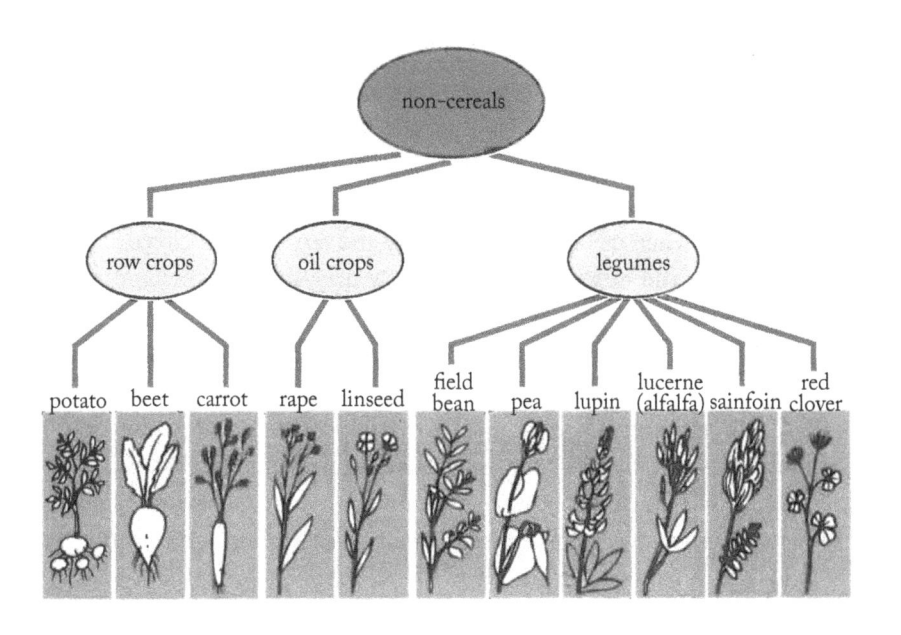

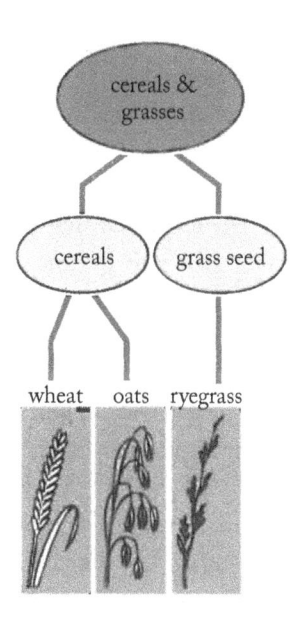

Figure 11. Major crops.

Non-cereals

These include row crops, oil crops and legumes (grown for pulses or for fodder).

Row crops are mainly potatoes, sugar beet, maize for silage and vegetables. They give high and extremely high yields but are demanding when it comes to soil quality and fertilization. Greedy for both humus and nitrogen they frequently have shallow roots and do little to help the crumb structure.

Most oil crops are members of the cabbage family, e.g. rape, mustard and oil radish. Linseed is a member of the flax family and is rarely grown as it is sensitive to competition from weeds. Sunflowers also count as oil crops. Rape, mustard and oil radish are important catch crops because they produce a large volume of green matter in a relatively short time. If the leaf mass is not sufficient for fodder, shallow ploughing in will help to build up the soil. They are generally deep-rooted, helping the crumb structure, and are nitrogen consumers and neutral with regard to humus.

Care must be taken to see that non-cereal crops include an adequate proportion of legumes.

The legumes that grow in temperate European latitudes belong to the pea family (Leguminosae). They have the advantage that their roots live in symbiosis with bacteria that fix atmospheric nitrogen, so that the soil nitrogen level is enhanced. Furthermore some species are very deep-rooted, prefer soils rich in lime and are able to make phosphorus and potassium accessible (to a depth of about a metre, sometimes a great deal more).

Pulses (field bean, pea, vetches, lentil, lupin, vetchling, soya bean) grown as green fodder leave the soil with a good crumb structure. They will use part of the stored nitrogen as they ripen. Grown for grain they have to be sown more thinly and progressively lose their foliage from below upwards, so that weeds become a late problem and crumb is reduced. Used for green manuring they are among the best preceding crops since they improve nitrogen, crumb and humus. Annual clovers – red, reversed (Persian), Egyptian, crimson and white – grown as a monocrop leave about 50 kg of nitrogen and 2 tonnes of root residues (dry matter) per hectare (45 lb and 1 ton/acre). Being deep-rooted, they open up the soil and improve crumb structure. The organic root mass decomposes relatively quickly as the C/N ratio is lower than with perennial leys. This means an increase in yield for the crop that follows immediately, but no or only short-term soil improvement. Perennial clover/lucerne/grass mixtures are the best type of preceding crop:

- about 100–180 kg/ha (90–160 lb/acre) of nitrogen is gained over two or three years,
- root residues make up 4–6 tonnes/ha (dry matter, 1½–2 tons/acre),
- crumb and soil structure are enhanced to an optimum degree.

The extended resting period helps the soil, and the humus level can be increased to lasting effect by mixing deep-rooted legumes (lucerne, red clover, sainfoin, white melilot, etc.) and shallow-rooted grasses with a large root mass (sheep's fescue on its own and a mixture of lucerne and false oatgrass each produce more than 10 tonnes/ha, 4 tons/acre, of dry root matter). Long-term leys are also important for weed control.

Within the group of grasses and cereals, the situation is approximately the same, but non-cereals show a major difference between row crops on the one hand and legumes on the other with regard to long-term preceding crop effect. The oil crops are about halfway between. It is evident, therefore, that account must be taken not only of the relation of non-cereal plants to cereals and grasses in deciding on a sequence but also of the ratio of row crops to legumes within the first group. Legumes may be grown as a catch crop to balance a high ratio of row crops where this is possible.

Catch crops

It is not always entirely feasible to maintain 'green cover' for arable land, and for example frost has a beneficial effect on heavy clays and loams that are ploughed and left open. However, an effort should be made to come closer to the ideal of green cover whenever the opportunity arises.

catch crop	following crop		
red clover and grass	1) potatoes 2) winter rye		
white clover and grass	1) sugar beet (damp site) 2) spring barley		
white melilot and grass	1) potatoes 2) winter rye		
serradella and grass	1) potatoes 2) winter rye		
sweet strain of fodder lupin and serradella	1) potatoes 2) winter rye		
winter rape and winter vetch	1) potatoes 2) spring rye		

0 50 60 70 80 90 100   10 20 30 %
catch crop   following crop   increase in yield

increase for 1st following crop ▨ increase for 2nd following crop

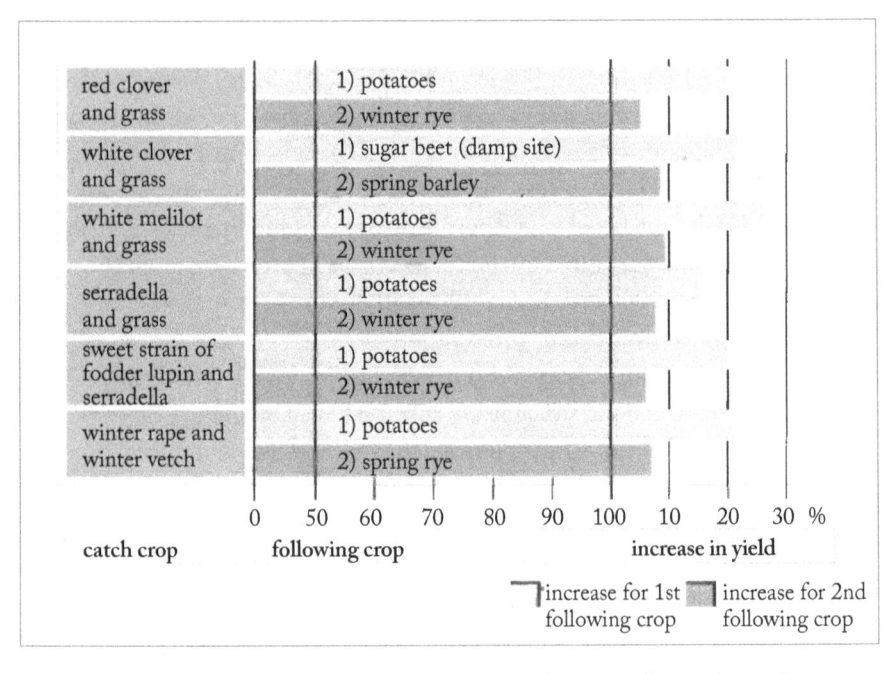

Figure 12. Preceding crop effect of leguminous catch crops on first and second following crops (Ruebensam & Rauhe 1964). Without catch crop = 100%.

Figure 12 shows the direct positive effect that legumes grown as a catch crop have on the first following crop, with the yield just on 20% higher, and on the second, with the yield about 5% higher compared to the control (100%).

Catch crops that produce sufficient aerial parts to serve as fodder should always be used for fodder.

Greenstuff has much greater value than fodder, with manure production also taken into account. Even without the manure effect, the graph in Figure 13 shows that the yield in grain equivalents of a lupin and serradella mixture grown as a stubble crop was 11% higher when used for fodder than for green manuring.

The application of composted manure to soil-building crops such as leys has a lasting effect on soil fertility.

It makes good sense to utilize the catch crop potential of suitable sites, particularly as this also helps to balance one-sidedness in cropping sequences.

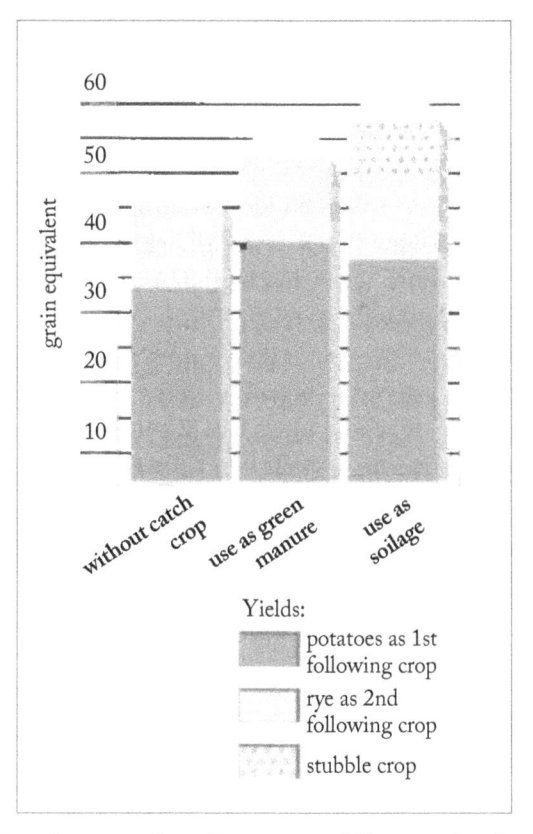

Figure 13. Preceding crop effect of legumes as stubble crop utilized as green manure or as fodder (Ruebensam & Rauhe 1964).

Manuring as part of the sequence

In biodynamic agriculture the principle is that 'manuring should enliven the soil' rather than feed the crops, and this means that the manuring system is different compared to conventional methods.

The use of composted manure is simpler than that of fresh or partly rotted solid and liquid dung as it permits flexible timing and there is no risk of burn when it is applied to crops. Nor does well-rotted manure suffer from exposure to sun or frost when used as a top dressing.

Soil development is encouraged by using compost in conjunction with crops that give lasting improvement in soil fertility. If Landsberg mixture (see page 190) is grown as an over-wintering catch crop preceding maize for silage, compost for the maize is actually applied in August, before the catch crop is sown. The mixture will leave the soil in an ideal condition for the maize crop. An application of dung liquor after drilling the maize, or between rows when the plants are in the juvenile stage, will be all that is needed to get a good yield.

The planned use of manure helps farmers to establish a fully developed individual farm organism. The effectiveness of soil-improving crops is enhanced by the use of composted manure. With two-year leys, the manure is applied for the second year. Good results are also seen with composted manure applied to catch crops. The row crops that follow usually give good average yields in the long term, without seriously affecting the humus and nitrogen balance of the soil.

The crumb layer is best deepened by ploughing in composted manure when a ley of several years is ploughed in summer. The beam should be set high and narrow for thorough mixing and aeration. Care must be taken to incorporate not more than one, or at most two, centimetres of subsoil in the topsoil at intervals of six to eight years. Root residues and compost will ensure lasting results with this operation.

Examples of cropping sequences

The principles discussed so far must be taken into account in designing the best possible cropping sequence. A good first step is to set out crops that in the given site make good pairs in a sequence. See Table 14.

When cereals follow one another, catch crops should be grown in between if at all possible. It appears that so far no accurate experimental data are available on cropping sequences in organic farming. Practical experience shows enormous variation at local and

field beans	– winter wheat	winter rape	– winter barley
peas	– winter cereal	(catch crop)	– winter wheat
lupins	– winter rye	winter barley	– vetch and rape
	– winter wheat		– Landsberg
	– potatoes		mixture
	– spring cereal	winter rye	– Landsberg
red clover	– winter wheat		mixture
	– spring wheat	early potatoes	– winter rape
	– oats		– winter barley
	– potatoes	med. early pot.	– winter rye
lucerne (alfalfa)	– spring wheat		– spelt*
	– oats	late potatoes	– winter wheat
	– sugar beet	potatoes	– carrots
	– potatoes		– beet
vetch and rape	– potatoes		– spring cereal
vetch and rye	– potatoes	carrots	– peas
	– maize for silage	maize for silage	– winter wheat
	– beet seedlings		– spring barley
Landsberg			– spring cereal
mixture	– maize for silage	fodder and	– winter wheat
	– beet seedlings	sugar beet	– spring cereal
	– marrowstem kale		

* Spelt is self-compatible and comes between winter rye and winter wheat for its qualities in cropping sequences.

Table 14. Useful pairs for cropping sequences.

farm level. In many cases accurate data are not available on yields and weather conditions. Table 15 provides a rough indication for the suitability of different sequences (Deutsch 1972).

	preceding crop		following crop	suitability
winter rye	winter wheat		winter rye	1
oats	spring barley	spring wheat		2
winter barley				3
oats			winter wheat	1
winter rye				2
spring barley	winter wheat	spring wheat		4
winter barley				5
winter rye			winter barley	1
spring wheat	oats			2
winter wheat				3
winter barley	spring barley			5
oats	winter rye		spring wheat	2
spring barley	winter barley	spring wheat		3
winter wheat				4
winter rye	winter wheat	spring wheat	spring barley	2
spring barley				3
winter barley	oats			5
winter rye	winter barley	winter wheat	oats	1
spring wheat				4
spring barley	oats			5
1 excellent; 2 excellent or good; 3 moderately good; 4 not good; 5 unsuitable.			

Table 15. Suitability of different cereal sequences.

6. CROPPING SEQUENCE

Example 1

	non-cereal 33% cereal & grass 67%	six fields N CC N CC	16.7%/field
1) CC	potatoes	sugar beet	peas*
2) CC	rye serradella	wheat rape/mustard	wheat rape/mustard
3) CC	spring barley red clover sown	spring barley red clover sown	spring barley red clover sown
4) CC	red clover*	red clover*	red clover*
5) CC	oats	oats	oats
6) CC	winter rye lupins**	winter wheat wh. clover/black medick**	winter rye serradella*
legumes			33%
main crop	16.7%	16.7%	16.7%
CC	33%	16.7%	
row crops	16.7%	16.7%	–
crucifer family	–	–	–
CC	–	16.7%	16.7%

* half quantity of composted manure, c. 18 tonnes/ha (7 tons/acre)
** full quantity of composted manure, c. 35 tonnes/ha (13 tons/acre)
N non-cereal crop; CC catch crop

Table 16. Six examples of cropping sequences to demonstrate the wide range of possibilities.

Example 2

	non-cereal 37.5% cereal & grass 62.5%	eight fields N CC N CC N C	12.5%/field
1) CC	potatoes	fodder beet	sugar beet
2) CC	winter rye serradella	winter wheat peas/vetches	spring wheat
3) CC	spring barley red clover sown	oats mustard*	winter rye red clover sown
4) CC	red clover	peas	red clover
5) CC	oats	winter wheat* wh. clover/black medick	oats
6) CC	winter rye Landsberg mixture**	spring barley red clover sown	winter rye Landsberg mixture**
7)	maize for silage white clover undersown	red clover	maize for silage CC white clover undersown
8) CC	spring barley lupins or ** white clover	winter rye lupins**	winter wheat peas/vetches**
legumes main crop CC *row crops* crucifer family CC	12.5% 50% 25% – –	25% 37.5% 12.5% – 12.5%	12.5% 37.5% 25% – –

* half quantity of composted manure, *c.* 18 tonnes/ha (7 tons/acre)
** full quantity of composted manure, *c.* 35 tonnes/ha (13 tons/acre)
N non-cereal crop; C cereal or grass for seed; CC catch crop

6. CROPPING SEQUENCE

Example 3

	non-cereal 40% cereal & grass 60%		five fields N CC N C	20%/field
1) CC	potatoes		sugar beet	field beans
2) CC	winter rye		winter wheat rape/mustard	winter wheat* peas/vetches
3) CC	winter barley* red clover sown		oats vetch and rye**	spring barley red clover sown
4) CC	red clover		maize for silage white clover	red clover
5) CC	oats/winter rye peas/vetches lupins**		spring barley peas/vetches**	oats rape/mustard*
legumes main crop CC *row crops* crucifer family CC		20% 20% 20% – –	– 60% 40% – 20%	40% 20% – – 20%

* half quantity of composted manure, *c.* 18 tonnes/ha (7 tons/acre)
** full quantity of composted manure, *c.* 35 tonnes/ha (13 tons/acre)
N non-cereal crop; C cereal or grass for seed; CC catch crop

69

Example 4

	non-cereal 50% cereal & grass 50%	six fields N C N C etc. N N CC N C		16.7%/field
1) CC	potatoes*	potatoes		ley
2) CC	winter rye Landsberg mixture**	carrots/fodder beet*		ley*
3) CC	maize for silage/beet seedlings peas/vetches	winter wheat* peas/vetches		winter wheat
4) CC	oats rape-mustard	spring barley red clover sown		oats rape/mustard*
5) CC	peas	red clover		field beans
6) CC	winter rye serradella*	winter wheat peas/vetches**		winter wheat* ley sown
legumes				
main crop		16.7%	16.7%	50%
CC		33.3%	33.3%	16.7%
row crops		33.3%	33.3%	–
crucifer family		–	–	–
CC		16.7%	–	16.7%

* half quantity of composted manure, c. 18 tonnes/ha (7 tons/acre)
** full quantity of composted manure, c. 35 tonnes/ha (13 tons/acre)
N non-cereal crop; C cereal or grass for seed; CC catch crop

6. CROPPING SEQUENCE

Example 5

non-cereal 62.5% cereal & grass 37.5%		eight fields N N C N N C N C	12.5%/field
1) ley CC		potatoes	ley
2) ley* CC		carrots/beet*	ley
3) winter wheat CC peas/vetches**		oats rape/mustard	winter rye
4) potatoes CC white clover**		peas	winter rape*
5) field beans CC lucerne (alfalfa) sown		winter wheat*	potatoes
6) winter wheat CC rape/mustard*		lucerne (alfalfa) rape/mustard	winter rye
7) peas CC		lucerne (alfalfa)	peas
8) winter barley* CC ley sown		spring wheat vetches/peas**	winter rye ley sown
legumes main crop CC *row crops* crucifer family CC	50% 12.5% 12.5% – 12.5%	37.5% 12.5% 25% – 12.5%	37.5% 12.5% 12.5% 12.5% 12.5%
* half quantity of composted manure, *c.* 18 tonnes/ha (7 tons/acre) ** full quantity of composted manure, *c.* 35 tonnes/ha (13 tons/acre) N non-cereal crop; C cereal or grass for seed; CC catch crop			

Example 6

	non-cereal 67% cereal & grass 33%		six fields N N N C N C	16.7%/field
1) CC	lucerne (alfalfa)		ley	red clover**
2) CC	lucerne (alfalfa)		ley**	potatoes
3) CC	lucerne (alfalfa)*		potatoes	carrots/beet*
4) CC	spring wheat rape/mustard*		winter rye rape/mustard	winter wheat
5) CC	peas		winter rape*	peas
6) CC	winter wheat* lucerne (alfalfa) sown		winter rye ley sown	winter barley* red clover sown
legumes main crop CC *row crops* crucifer family CC		67% – – – 16.7%	33% – 16.7% 16.7% –	33% – 33% – 16.7%

* half quantity of composted manure, *c.* 18 tonnes/ha (7 tons/acre)
** full quantity of composted manure, *c.* 35 tonnes/ha (13 tons/acre)
N non-cereal crop; C cereal or grass for seed; CC catch crop

7. Seeds and Sowing Techniques

Next to a good seed bed and the application of manure, the inherent quality and external features of the seed are important factors in achieving good yields. In the past, numbers of local ('land race') varieties well adapted to local conditions were successfully grown for long periods. They suited the geographical and soil conditions and also met the special food requirements of the local population.

Years of work in modern plant breeding have produced highly selected strains that give high yields in conjunction with intensive applications of mineral fertilizers. They are, however, more susceptible to fungus attack, pests and extreme weather conditions than the older varieties, necessitating extensive use of plant protectives. Degenerative changes occur when such seed is grown on through successive generations. Marked reductions in yield mean that new seed has to be used after two years; this is the case with cereals, particularly rye, which is a cross-pollinating crop. In Germany, approximately 50% of cereal seed is bought in as certified seed. The advice is actually to buy 100% new seed every year. Certified seed is produced from approved basic seed obtained directly from the breeder; it is not intended for further seed production and must meet a number of legal and other requirements and derive from listed varieties.

The aim and purpose of biodynamic farming methods – to develop a farm that is as far as possible an independent entity – means a totally different attitude with regard to seed provision.

Choice of variety

The first step is to find the right variety. Careful note taken of trials run in different localities by the agricultural advisory centres points us in the right direction. The problem is that large amounts of readily soluble mineral fertilizers are used in these trials, so that the results may be misleading.

A variety has to be grown for a number of years before useful information on its inherent qualities is obtained. Long-term fertility depends very much on resistance to poor growing conditions and adaptability to variable growing conditions These two qualities indicate the inherent vitality of the variety.

Inspection walks are a regular feature in the work of regional biodynamic farming groups. The observations made on these walks often permit reliable predictions to be made as to the usefulness of individual varieties. The final decision will depend on trials done on one's own farm for a minimum of three years. Once a particular variety has proved suitable, every effort should be made to keep it sound and productive for many years on one's own farm. Trials done in Witzenhausen (near Kassel, Germany) have shown that both old and highly selected varieties may prove suitable.

5% or more above mean	Mean yield (± 5%)	5% or more below mean
1 Vuka	9 *Carsten 6*	20 *Stir. Farino*
2 Disponent	10 M. Huntsmen	21 *Breustedt Werla*
3 Saturn	11 Kormoran	22 *Derenburger Silber*
4 Okapi	12 Diplomat	23 *Baulaender Spelz*
5 *Merlin*	13 *Heine 110*	24 *Hauter 2*
6 Caribo	14 *Peragis*	25 *Hess. Landsorte*
7 *Carsten 2*	15 Ural	
8 *Heine 4*	16 *Pfeuffer Schernauer*	
	17 *Rimp. Bastard*	
	18 *Sv. Kronen*	
	19 *Carsten 5*	
Names given in *italics* are old varieties.		

Table 17. Comparative yields of 9 highly selected modern varieties of winter wheat, 15 old local varieties and 1 spelt. Site: northern Hesse, biodynamic farm. Mean figures for 3 years' grain yields (ranking order). Manuring: c. 15 kg org. N/ha (13 lb/acre) and year (dung liquor) (Padel 1984).

High yield capacity depends on the plants' assimilation and mineral uptake capacities (leaf and root development). Modern varieties frequently have the two uppermost leaves especially broad, long and robust. Light utilization is improved by having the leaf more horizontal to the stem.

The effectiveness of horn silica is partly due to improved assimilation; horn manure and biodynamic seed baths encourage root development. Observations made at Talhof Farm have shown that following the application of horn silica the leaves are at a good angle to the light, exposing as much of their surface area as possible as they follow the movement of the sun.

Trials with new varieties that had shown particularly well-developed upper leaves in variety tests gave good results at the Talhof Farm. 'Adler' is a spring wheat variety that has given good yields and quality and stayed in good health on the farm for sixteen years. The period of extreme dryness in 1976 caused premature ripening of this crop. After much thought, the miserable, shrivelled grain, which however had good germinating power, was used for seed and produced healthy crops with good yields. 'Nackta' is a naked barley grown successfully for fifteen years. A spring wheat variety that a breeder had originally supplied for trials has now been grown for nine years.

Growing quality seed

Some indoor and outdoor gardeners are said to have 'green fingers', for their plants always do well. In the days when fields were still sown by hand, some farmers always had particularly strong and healthy crops. Neighbours would seek them out and ask them to sow their fields for them.

Farmers who wish to produce their own seed and keep it healthy and vital for long periods need to develop special skills. These include the faculty of careful observation and the ability to judge the quality of crops and individual plants. We need to familiarize ourselves with the growth rhythms and special needs of crops and their companion herbs and grasses. It means that farmers have to develop an inner relationship to the plant world, and the same of course also holds true for the breeding and keeping of animals.

Sowing times

It is immaterial to a plant whether it is grown for food, for fodder or for seed. For the fruit, however, the distinction has to be made, for it certainly matters if it is to have more nutritive or reproductive power.

Cultivated plants differ from wild species mainly in that they provide more nutrients, though this is at the cost of their ability to reproduce themselves.

Trials suggested by Rudolf Steiner and done at the Pilgramshain estate in Silesia have shown that cereals sown close to the winter months produce seed with high reproductive power. Sowing closer to the summer season tends to enhance the nutritive value.

This relationship to the seasons can be utilized by sowing winter crops later and spring crops earlier to obtain healthy seed with high growth and yield performance. Timely and thorough preparation of fields to be sown close to the winter season will go even further in assuring good results. Nutritive quality is enhanced by tilling the soil and sowing food and fodder crops closer to the summer months, i.e. earlier for winter crops and later for spring crops.

First sequence	Second sequence
1st year: rye 2nd year: ley 3rd year: ley 4th year: potatoes 5th year: wheat 6th year: oats 7th year: peas 1st year: rye	1st year: rye 2nd year: red clover 3rd year: potatoes 1st year: rye
The first cropping sequence includes five crops (and a two-year ley) sown in six years. It is the better one for growing rye seed. In the 2nd sequence the interval between growing rye is only two years.	

Table 18. Selection of field for seed production as part of cropping sequence, using rye as an example.

Figure 14. Part of the field suitable for collecting seed.

Experience will show which are the best sowing times for different crops on a particular site. The shift towards winter or summer should not differ from these times by more than three months, or yields may be reduced.

When plants have grown for seed close to winter for two years, the next sowing must be at the 'normal' time, or even a little closer to summer, to avoid one-sided development. As a precaution, save some seed for the following autumn to avoid losing a variety that has proved successful on the farm; this can be used if the new harvest has poor germinating power and does not produce useful seed.

Choice of growing site

Crops grown for seed are best included in the existing cropping sequence. If several sequences are used on the farm, chose the field where the interval for the same crop is the longest.

Parts of fields where the soil is in extremely rich or poor condition are not suitable, nor are headlands and marginal strips. This leaves a sufficiently wide strip down the middle of a field for growing for seed, with the rest of the field used to grow the same crop for food or fodder in the usual way. Marker posts and good seed bed preparation allow drilling close to the winter season even if there is light snow cover, taking account of relevant astronomical data (see *Maria Thun Biodynamic Calendar*). The soil should not be frozen, however. For spring crops, prepare the chosen site in autumn.

Areas selected for seed growing are either not fertilized at all or given just a small amount of well rotted or humified manure (12–16 tonnes/ha, 4½–6 tons/acre); they should not be given rich manures that would force growth.

In-field selection

Having chosen the sowing time and the field, another operation to ensure healthy and vital seed is in-field selection. 'Positive mass selection' consists in selecting strong healthy specimens spaced at regular intervals, i.e. with no gaps between them nor as solitary plants. Ears and panicles should be filled with well-developed grain, including base and tip. Preselection is done during the active growth period, taking special account of leaf and culm development and using stakes to mark selected specimens. Final selection based on the above criteria is made shortly before the whole field is harvested. The selected ears and panicles are cut off below the uppermost node in the case of rye, wheat and oats, and at the second highest node in the case of barley. This material provides the basic seed; it must be dried with care and stored in a separate place. It is then grown for seed in a selected field during the next season and provides the basis for subsequent food or fodder crops.

It does need perseverance to acquire the knowledge and skills needed for in-field selection and not every farmer is prepared to put this kind of effort into maintaining and enhancing seed quality. A simpler method that is reasonably effective is 'negative mass selection': carefully remove all abnormal and atypical specimens and all plants of different species from the field selected for seed growing and use a combine harvester to harvest the seed crop separately from the food crop.

An additional measure is to sort the machine-cleaned seed by hand every three or four years during the winter, discarding all abnormal, damaged and discoloured seed as well as seed from other species.

This is a labour that has to be done. It is good to remember that such intense involvement with plant nature will develop completely new relationships, capabilities and powers.

Within the regional group, individual farm communities may each take responsibility for a particular crop in which they have a special interest and which suits their particular site. Any form of plant breeding that goes beyond this exceeds the limits of practical farming and calls for special capacities and equipment.

Seed baths

In conventional farming, seed is usually treated with more or less toxic dressings to protect it from pests and fungus attack.

Martha Künzel and Franz Lippert have done trials for many years and developed a seed treatment that strengthens the vitality and resistance of plants. This involves seed baths using selected biodynamic preparations for different plants. Plant growth is enhanced as a result, germination tends to be faster, no doubt partly due to moisture uptake. Legumes have been found to produce greater numbers of nodules. These result from the activities of nitrogen-fixing bacteria living in symbiosis with leguminous plants. Growth is improved and compared to untreated controls these plants prove healthier and give better yields. They are able to cope better with poor weather conditions and develop a stronger and more extensive root system.

Seed and plant material is treated with the appropriate extracts the day before sowing. Table 19 gives details of seed baths for different seeds.

horn manure	Stir for 1 hour immediately before use. Use for spinach.
valerian preparation	Stir for 15 minutes before use, a tablespoonful to 10 litres of warm water. Use for wheat, fodder beet, sugar beet, leeks, onions, tomatoes, celeriac, potatoes
birch pit concentrate	1 part + 4 parts rain water + 5 parts whole milk; stir vigorously for 5 minutes, leave to stand for 20–24 hours, stir another 5 minutes before use. Use for fodder and sugar beet, carrots, potatoes (made up without milk).
yarrow preparation	Stir 1 portion (c. 1–2 ml) in 3 litres of hand-warm rain water vigorously for about 5 minutes, leave to stand for 20–24 hours (an occasional stir helps); stir briefly before use. Use for rye, grass seed.
wild camomile prep.	As for yarrow. Use for legumes, radishes, rape, mustard, tulips, cabbage varieties.
nettle preparation	As for yarrow. Use for barley.
oak bark preparation	As for yarrow. Use for oats, lettuce, potatoes, dahlias.

Table 19. Seed baths (Künzel 1949).

Cereal and legume seeds will need two to three litres per 100 kg (1–1½ quarts/100 lb). Using a hand-brush, or a knapsack sprayer for larger amounts, spray the mound of seeds, turning it three times with a shovel in the process. Cover with sacks and leave for two hours. The small amount of moisture will distribute itself evenly through the heap in this time. Spread it out to dry, so that there will be no problems with the drilling machine, and turn once more if necessary.

> Seed grown on the farm for many years must be kept clean. Foreign elements of any kind, weed seeds as well as seed from other cultivars, are deleterious.

Put smaller amounts into fabric bags and suspend in the liquid for 15–20 minutes; dry in a shady place before sowing. Seeds of row crops require two hours' soaking.

Seed potatoes are also sprayed, turning the heap at the same time, but this is done three times in two weeks. Birch pit concentrate will improve the quantitative yield, whilst valerian gives better resistance to degenerative diseases and late blight *(Phytophthora infestans)*.

Carrot seed is best dried in a spin dryer after treatment and sown immediately. All bathed seed and plant material must be sown or planted within two days, which means that the operation cannot be done in advance to utilize slack times.

Seed bath treatment gives farmers opportunity to give the powers of their heart to the seed, letting a stream of forces enter into plant development.

External qualities of seed

Size, weight, shape

All effort will be in vain if the seed does not meet certain external quality standards. The size and shape of seed grains should be typical for the species and variety. Overlarge grains frequently come from jagged ears that are not full or from single plants grown in a part of the field where conditions were too favourable, e.g. due to overlap of manure and dung liquor applications. The grain size should always be barely above average. Grains that are full and round are given preference; they are fully ripe and of the right weight.

Clean seed

Care must be taken to see that seed grown on the farm for many years is clean. Foreign elements of any kind, weed seeds as well as seed from other crops, are deleterious. Just a few grains of rye can seriously contaminate wheat seed.

Utter cleanliness is therefore essential at every stage:

- Sacks must be thoroughly clean.
- Broadcasting machines and seed drills must be free from grain residues, with tubes, disks and every nook and cranny scrupulously clean.
- Where fields are immediately adjacent, with no margin between, close the shutter for the outermost drill tube on the appropriate side.
- Combines and other machines used for harvesting have to be carefully cleaned. Special care must be taken when threshing is done on the edge of a field, to avoid inclusion of other seed.
- All conveyor equipment such as pneumatic grain conveyors, augers and elevators, and the seed cleaner, must be kept clean.
- Great care is also needed in the seed store.

With bought-in seed, especially fodder plant seed, great care must be taken to see that no seeds of undesirable persistent weeds are present, for it is in this way that farms, and particularly farms growing a high percentage of fodder crops, are fouled with curled and broad-leaved dock (*Rumex crispus* and *R. obtusifolius*). Prevention consists in using farm-produced seed of clover species again, with all weeds carefully removed from crops intended for seed production.

Germinating power

Reduced germinating power may be due to a vast number of reasons and it is advisable to do a germination test on farm-produced seed. Spread a sample of 100 grains between two sheets of wet blotting paper on a plate. The method may be used with cereals, fodder vetch, rape, cabbage and clover species. Large seeds such as maize, field bean, beet and pea are pushed into moist sand, keeping the top of the seed level with the surface.

Points to be considered when doing germination tests:
- Keep moist but not wet (lack of oxygen) at all times.
- Light conditions do not matter with agricultural seeds such as cereals, pulses, beets, cabbage and its allies, rape, etc., as they are neutral to light. Only grasses need light for germination, so that the sample should not be left in the dark. Horticultural species often need darkness to germinate (onion, cucumber, tomato, etc.).
- Room temperature is normally adequate. Alternating temperatures accelerate germination.

Germination tests for cereals may have to be done at low temperatures (8–12°C, 46–54°F) if seeds have not ripened by autumn.

After ten days, or for grasses 14–21 days, depending on species, count all seeds that have germinated properly, discarding any that show abnormalities such as glassiness, twisted sprouts, fractures or mouldiness and seeds that have merely swelled up.

The number of properly germinated seeds is equal to the percentage germinating power.

Water content

Water content is a major factor in the keeping quality of seeds. Cereals containing 16–20% of water must not be stored in sacks; above 20% the grains swell up. Permissible levels have been given by Huebner (1955):
- cereals and pulses 15.0%
- rape and related species 12.0%
- clover 12.5%
- grass 14.5%
- vetches 14.0%

Table 20. *Quality criteria for seed.*

Crop	thousand seed weight g	germinating power (approved seed) %	viability years	weight per hectolitre kg	minimum germinating temperature °C	sowing depth cm
wheat		94	3–4	65–84	2–4	2–4
large grain	50					
small grain	40–50					
barley		94	2–3	60–65	2–4	2–4
large grain	50*1					
small grain	40–45*1					
rye		94	1–2	60–80	2–4	2–4
large grain	35					
small grain	30–35					
oats		94	2–3	35–56 (50*2)	2–4	2–4
sorted	35–45					
unsorted	25–35					
maize		85*3–90	2–3		8(10)	4–8
large grain	400	85				
small grain	100–250	85				
millet	2.2–5.4*4	85	3–4	60–70	6–10	1.5–2
sorghum	8–14*4	85	3–4		10–12	2–2.5
rape			3–4		2–3	2–3
winter	4.0–6.5	94				
spring	2.5–4.5	90			2–3	1–2.5
turnip rape			3–4		1–3	2–3
winter	2.0–4	94				
spring	2.0–3.5	90			1–3	1–2.5
flax for fibre	3.4–5.3	90	2		2–4	1–2
flax for oil	5.4–14	80*5	2		2–4	1–2
flax for oil & fibre	6.0–8.0	80*5	2		2–4	1–2
white lupin	340–520	80*5	poor		4–5	3–4
blue lupin	150–200	94	poor		4–5	3–4
yellow lupin	110–200		poor		4–5	3–4
field bean			3–4		3–4 (9)	5–8
large seeded	600					
medium seeded	500–600					
small seeded	350–500	88				
dwarf bean	250–600	94	2–3		8–10*6	5–6
vetchling	250–400		3–4		3–4	3–4
green pea	150–450		3–4		1–2*7	6–8*7
field pea	100–200	88	3–4		1–2	2–5
soya bean	80–200				8–10	4–7
common vetch	20–140	88 (83)*8	3–4		1–2	2–3
lentil	25–70	92			5–8	2–3
hairy vetch	25–40		3–4		3–4	1.5–2.5

Crop	thousand seed weight g	germinating power (approved seed) %	viability years	weight per hectolitre kg	minimum germinating temperature °C	sowing depth cm
red clover	1.5–2.5	85*5	1–2	c. 75	0–5	1–2
white clover	0.65–0.86	85*5	2	76–80	0–5	1–2
Alsike clover	0.6–0.7	85*5	1	75	0–5	1–2
Egyptian clover	2.6–3.2		1		2–6	1–2
lucerne (alfalfa)	2.0–2.5	85	3	c. 77	1–6	1–2
white melilot	1.8–2.0	80*)	1		1–6	1–2.5
melilot	1.7–2.0				1–6	1–2.5
crimson clover	3–4	85	1	76–80	3–8	2–3
black medick	1.2–2.0	85*5	1	76	1–6	1–2
birdsfoot trefoil	1.2	75*5	4–5	70–78	1–6	1–2
greater birdsfoot trefoil	0.5	75*5	–	78	2–6	2–3
serradella	4	75	1–2		3–8	4–5
sainfoin					1–6	4–5
whole pod	20–25	80	3–6	33		
seed	12–15		1–2			
kidney vetch	2.4	75	1		0.5	2–3
awnless brome grass	4.0		3–4			
false oatgrass	3.2	80	2–3			
perennial ryegrass	2.0	90*9	2–3			
Italian ryegrass	2.1		1 (2)			
hybrid ryegrass	2.3		2			
annual ryegrass	2.5		1			
meadow fescue	2.1	86	2–3			
creeping fescue	1.1	86	3			
cocksfoot	1.0	85	2–3			1–2
meadow foxtail	0.7–0.8	70	3			
reed grass	0.5–0.9	75	2–3			
yellow oatgrass	0.2–0.4	70	3			
timothy	0.4–0.5	90	2–3			
smooth meadow grass	0.4	80	3–4			
swamp meadow grass	0.2	(75–80)*10	3–4			
fiorin	0.05–0.08	85	3			
dogstail	0.5	80	2–3			

*1 Spring barley: down to less than 25 g acceptable; winter barley, down to 30–35 g.
*2 Seed quality
*3 Germinating power sometimes only 80%, due to high water content (20–35% or more) at harvest. Careful drying increases germinating power to 95% or more.
*4 For absolutely dry seed.
*5 Low because of hard seed coat – see Table 21.
*6 *Phaseolus* species
*7 *Pisum* species
*8 spring vetch (winter vetch)
*9 ryegrasses
*10 *Poa* species

Table 20 gives a comprehensive overview over the most important quality criteria, minimum germination temperatures and depth of sowing for a number of agricultural crops.

lupin	high (highest for yellow lupin)
red clover	6–10%
white clover	10–25%
Alsike clover	5–20%
melilot	10–15%
black medick	generally low, but may be up to 25%
birdsfoot trefoil	20–40%
greater birdsfoot trefoil	10%

Table 21. Germinating power reduced by hard seed coat: percentage of hard seed (Huebner 1955).

Sowing techniques

Density

Crops have to have a certain density in a particular site if optimum yields are to be achieved. With cereals, for instance, these depend on the number of eared culms per square metre. It is well worth the effort to assess different varieties for the number of eared culms per germinated seed. It is a good idea to make a wooden or light-metal frame that is a metre square and place this on the crop. The number of plants, or of eared culms per square metre is then quickly counted. By keeping a record of the various counts for a number of years it is possible to determine the ideal number of plants per square metre or hectare for a given site. The amount of seed required is then easy to work out from

- the desired number of plants per square metre,
- thousand seed weight,
- germinating power,
- losses (%) (also determined on site),
- tillering (%).

Hoeing of cereals will mean 15% losses on average.

Seed required (kg/ha) = $\dfrac{\text{No. of plants/m}^2 \times \text{thousand seed weight (g)}}{\text{Germinating power} - \text{losses} + \text{tillering} - \text{hoeing loss}}$

For example for winter wheat (no hoeing):

Seed required = $\dfrac{400 \times 49}{95 - 10 + 15 - 0}$ = 196 kg/ha

---

Depth

The sowing depth depends on grain size, soil condition, climate and weather.

Large seeds need more moisture to swell and germinate and are more likely to find this in deeper soil layers, which is why they are sown more deeply. They also have greater powers of penetration, so that they are able to come up to the surface from below. The depth should be such, however, that sufficient oxygen is still available for germination. Seeds should be sown less deeply in wet and heavy soils with poor aeration than in well-aerated loose, light soils.

In areas where the climate is wet, or in wet weather conditions, it will also be necessary to sow less deeply than under dry conditions. It is generally best to sow as shallowly as conditions permit, as this will mean faster germination and an earlier start to assimilation. Plants that come up quickly are able to compete better against weeds.

seed in kg/ha	140	200	230
seed in g/m²	14	20	23
no of seeds/m² (mean thousand seed weight = 45 g)	311	444	511
seeds capable of germination/m² (95% theoretical)	295	422	485
plants/m² (theoretical 90%; normal 85–95%)	266	380	437
crop density = ears/m²	306	437	503
(theoretical 15%, normal range 10–20%)	293–319	418–456	480–524
crop density = ears/m²	260	372	428
with 15% loss due to hoeing	249–271	355–388	408–445

Table 22. Crop density calculated for winter wheat (Goetz & Konrad 1978).

Distance between rows

This will depend on the space required by the cultivar and on necessary cultivation operations (hoeing, harrowing, weed harrowing). Plants grow best if each is given approximately the same space. So far, however, no sowing technique has been evolved to achieve this. Closer rows mean a greater distance between plants in the row. Reducing the space between rows does get one closer to the ideal of sowing at equal spaces. Plants will then cover the ground more quickly and suppress weeds.

With cereals, a distance of 12–20 cm (5–8 in) between rows has proved effective, leaving tram lines to suit the vehicles to be used. If the tractor leaves very wide wheel marks, two rows may be sown close together next to the tram lines (Figure 15).

Crops grown on good growing soil or in well-manured fields will do well if sown in rows that are further apart. There will be fewer ears but the yield per hectare may well be increased as the ears are fuller and individual grains heavier.

7. SEEDS AND SOWING TECHNIQUES

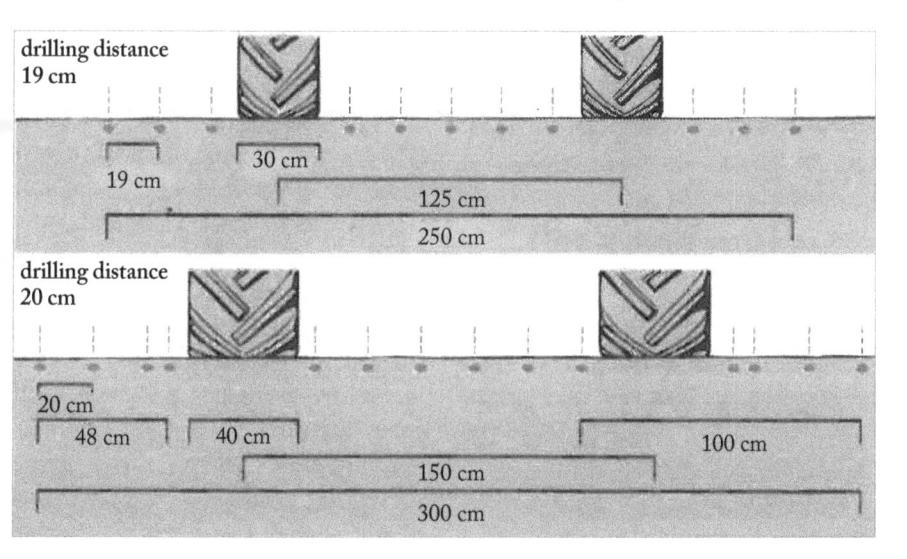

Figure 15. Distance between rows for cereals allowing for different widths of tram lines to permit hoeing.

Seed bed preparation and sowing

A well-settled soil with a top layer of fine crumbly tilth is the goal of seed bed preparation. The top layer needs to be coarser for larger and coarser seed grains and have a finer crumb for small seed.

Growth is assisted by sowing in a north–south direction, as this gives better warmth and light conditions. This is the method of choice where the situation permits. On sloping ground follow the contour lines to prevent or at least reduce soil erosion.

Modern drills are designed to deposit seed at regular intervals and at the required depth. Implements may be fitted with rollers to ensure contact with soil moisture for the seed. Coverers serve to loosen and spread a fine layer of earth over the seed, thus reducing evaporation. Compared to rolling the whole surface, this method reduces water loss to a much greater extent and is particularly valuable when drilling beet, maize and fine seeds such as carrot, rape, cabbage and its allies and onion. It may also prove useful when drilling wheat. Young wheat plants are somewhat better

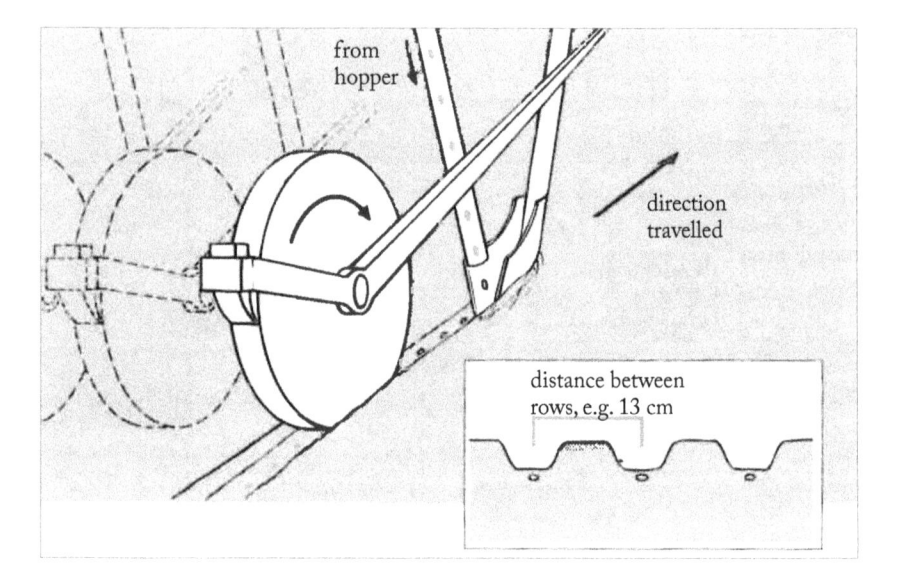

Figure 16. Effect of roller.

protected from frost in the slightly deeper grooves, with snow remaining in the grooves even in strong winds and protecting them.

The soil should have a moderate degree of moisture when drilling operations are carried out. This will also enhance the effect of horn manure applied to encourage seedling and root development. Appropriate sowing dates should be adhered to whenever possible.

8. Post Drilling Cultivations

The responsibilities of ecologically conscious farmers extend beyond the growing of crops to all the plants that are part of the local ecosystem.

On the one hand it is necessary to create conditions in which sufficient produce can be grown that meets high standards regarding taste, nutritive value and keeping qualities. On the other hand wild plants, too, are life forms that have their definite place and value in the ecosystem even if they cannot be said to have immediate economic value. Yarrow, wild camomile (scented mayweed), nettle, dandelion, valerian and oak, for instance, are needed to produce the compost preparations. Others benefit field crops merely by their presence, typical examples being sainfoin and cornflower in cereal crops.

Post drilling cultivations are designed to create and maintain the best possible conditions for healthy growth. Other important factors are crop density, a well-planned cropping sequence and good manuring practice. If a field is well prepared and in good tilth when drilling takes place, cultivation operations will keep it in that state until the crop itself provides the ground cover needed to protect the soil from the inclemencies of the weather. The combined effects of fibrous roots, soil organisms and ground cover produce a stable crumb that provides the best growing conditions. Cultivations also include weed control.

The agricultural system, site conditions and climate give rise to a specific accompanying flora. When flax was no longer grown, for example, the plant community that went with its cultivation (flax darnel, *Lolium remotum*; Linetalia or flax-associated flora) disappeared

(Knapp 1971). Increased cultivation of maize is favouring a special flora consisting of millet species, lesser bindweed and couch grass that is difficult to control. Farmers and the work they do clearly have a major effect on the natural environment.

Natural associations have been seriously weakened if not destroyed, which is evident from the fact that beekeeping is no longer possible in some agricultural areas in Central Europe. The rich variety of herbs and flowering plants has been banished from the fields, leaving only rape. And what good is it to partridges to be listed as an endangered species when they can no longer find the food and the cover they need? This impoverishment of the environment reflects the inner life of modern society, where economic values are so much to the fore that humanity fails to maintain the very basis of existence both in the physical world and in regard to the spirit.

> Schedule 8 of the Wildlife and Countryside Act of 1981 (UK) lists protected plant species in the United Kingdom.

Condition of soil	important for grassland	arable land
poor drainage of top and subsoil		
common horsetail (*Equisetum arvense*)		x
corn mint (*Mentha arvensis*)		x
coltsfoot (*Tussilago farfara*)	x	x
marsh woundwort (*Stachys palustris*)	x	
common reed (*Phragmites australis*)	x	x
poor structure and low concentration of lime		
pennycress (*Thlaspi arvense*)		x
common sorrel (*Rumex acetosa*)	x	x
curled dock (*Rumex crispus*)	x	x
broad-leaved dock (*Rumex obtusifolius*)	x	x
heartsease, wild pansy (*Viola tricolor*)		x
hare's foot clover (*Trifolium arvense*)	x	

8. POST DRILLING CULTIVATIONS

Condition of soil	important for grassland	arable land
corn spurrey (*Spergula arvensis*)		×
smooth finger-grass (*Digitaria ischaemum*)		×
meadow buttercup (*Ranunculus acris*)	×	
wild camomile or scented mayweed (*Chamomilla recutita*)		×
lack of stable crumb structure		
creeping thistle (*Cirsium arvense*)		×
knotgrass species (*Polygonum sp.*)	×	×
scentless mayweed (*Matricaria perforata*)		×
comfrey (*Symphytum officinale*)	×	
wind bent grass (*Apera spica-venti*)		×
good crumb structure for row crops with good supply of nitrates		
black nightshade (*Solanum nigrum*)		×
annual mercury (*Mercurialis annua*)		×
petty spurge (*Euphorbia peplus*)		×
annual nettle (*Urtica urens*)		×
common chickweed (*Stellaria media*)		×
stable crumb structure and neutral reaction		
lesser bindweed (*Convolvulus arvensis*)		×
scarlet pimpernel (*Anagallis arvensis*)		×
annual mercury (*Mercurialis annua*)		×
forking larkspur (*Consolida regalis*)		×
common toadflax (*Linaria vulgaris*)		×
white campion (*Silene latifolia*)	×	
dandelion (*Taraxacum officinalis*)	×	
yellow vetchling (*Lathyrus aphaca*)	×	
meadow clary (*Salvia pratensis*)	×	
summer pheasant's eye (*Adonis aestivalis*)	×	
chicory (*Cichonum intybus*)	×	

Table 23. Indicator plants

Available methods

Farms run on biological principles will usually have balanced cropping sequences with a good proportion of fodder crops. Undesirable herbs and grasses in arable and grassland provide an indication of site conditions. Knowledge of such 'indicator plants' is therefore considered important in ecological farming practice as they enable farmers to draw conclusions as to local soil conditions if they appear in large numbers. Table 23 gives details based on the work of Boas, Ellenberg and others (Appel 1979).

Another way is to classify such plants according to specific aspects. For instance, the appearance of weeds such as chickweed, deadnettle, wild camomile (scented mayweed), wind bent grass, slender foxtail or cornflower relates to weather conditions. Weeds common in fields used for spring crops are corn marigold, hemp nettles, charlock, wild radish, knotgrass and scarlet pimpernel. Orache, fat hen, gallant soldier, black nightshade and small nettle are typical associates of row crops. Other important categories are weeds reproduced by seed or root and those that need frost, light or darkness to germinate.

The presence of indicator plants cannot always be directly related to specific cultivation operations. Wild camomile or scented mayweed (*Chamomilla recutita* or *Matricaria chamomilla*) for example is considered indicative of lime deficiency. Excessive spread of this plant may however also be due to other causes and these need to be treated as well as applying lime. Wild camomile germinates in the light and tends to spread in areas of low crumb stability where the soil is inclined to puddle and encrust. Consolidation of the soil at middle level may also be responsible. Mulching and biological measures will overcome structural weaknesses of this kind.

A loose crumb structure may be maintained by loosening the soil, shallow digging in of compost and continuous use of weed harrow and roller, which prevents wild camomile from germinating.

Massive occurrences of coltsfoot, horsetail and corn mint indicate poor drainage in both crumb and subsoil. The condition may be improved by draining fields and grasslands. Useful methods are to put in effective drains, remove water by means of open drainage ditches and to create a herringbone ridge and furrow system for surface drainage, possibly combined with the creation of small ponds. As the water is drawn off, air penetrates the soil, supplying vital oxygen. The result will be a large increase in the microbe population and improved soil structure.

The next step is systematic fertilization with well-rotted farmyard manure and fermented dung liquor or slurries. The organic matter vitalizes the soil, which encourages the activity of soil organisms. Weed seeds are also more likely to be attacked by the increased fungus and bacterial population and this, possibly coupled with the production of inhibitory compounds, reduces their germinating power and vitality. Manures should be kept as free from weed seeds as possible, which is done by keeping heaps well covered and controlling humification by maintaining temperature and moisture levels. This may sometimes necessitate turning the heap. If there is still any doubt as to safety and quality, stick to the well-established rule: manures containing a high proportion of weed material should go on to permanent grassland, not on to arable land.

> Manure and compost containing a high proportion of weeds should go on grassland rather than arable land

Cropping sequences

Even the best planned cropping sequence will do little for crops that are thin or show gaps. This applies not only to winter wheat in spring, but also to red clover in summer. Couch grass and thistles will spread rapidly under those conditions. Defective crops should not be left to themselves, hoping that they will grow stronger and suppress the weeds. Inaction results in reduced yields and an increasing volume of weed seeds, with the rhizomes of perennial weeds growing on unimpeded. Undersown clover will help to some extent by providing ground cover. The advantages of planned cropping sequences have already been considered. Individual species differ in competitiveness, and one factor to be considered in designing cropping sequences is the balanced combination of 'weed promoters' and good 'weed suppressors'. Spring barley belongs to the first category; little straw and a short period of active growth mean that the soil is given little shade and hardly ever has complete cover. This offers plenty of opportunity for undesirable weeds to develop. The opposite is the case with winter rye, which does not even allow creeping thistles to come up.

Gaps in red clover due to the depredations of mice, clover crown and stem rot or winter kill help the spread of thistles and couch grass. One way of dealing with this is to use the more vigorous tetraploid red clover which is less susceptible to clover rot and gives better ground cover. Fields badly fouled with thistles were completely cleared by growing tetraploid red clover, and the wheat that followed was completely thistle free. Frequent cutting of long-term leys seriously upsets the growth rhythms of perennial weeds. Such leys should be planned for at least two years of use; three years of lucerne and grass are even more certain to give results.

Densely grown crops quickly followed by catch crops are excellent for weed suppression.

If such a time interval is not available and growing conditions are good, one summer crop of reversed or Persian clover will give almost the same result. To begin with, the clover can hardly be seen among the annual weeds, but after the first cut it will spread and suppress the germination of undesirable plants. Thistles do not tolerate four or five successive cuts. If necessary, turning the soil to some depth will suppress them further. The effect of under-sowing winter crops with white clover and black medick tends to be unreliable. Dense cereal crops (more than 4.5 tonnes/ha, 1.5 tons/acre) seriously inhibit their growth, with the result that there are bare areas in autumn where more weeds will grow. Recent developments have brought some improvement. A return to the use of mechanical hoes and hence an increase in the distance between rows to 17–20 cm (7–8 in) helps the clover and black medick to establish better. Seeds from weeds such as

8. POST DRILLING CULTIVATIONS

wild oats cannot reach the soil and are caught up in the dense 'clover pelt' where they perish. It is remarkable how quickly bothersome weeds can be controlled in this way.

Fodder mixtures of rape or mustard, field beans, vetches and peas as a catch crop will also contribute much to weed control if drilled in good time by the end of August or early September and if the weather is favourable. The weed seeds will germinate at the same time but be unable to compete with the vigorous growth of mustard and/or rape. The mixture also helps to suppress perennial weeds.

If sufficient nitrogen is available, dense stubble-sown crops of phacelia *(Phacelia tanacetifolia)* will rapidly form a dense carpet, inhibiting and killing off all other plants. Cattle do, however, take some time to get used to them. If the plant comes into flower it provides welcome 'autumn pasture' for bees. Phacelia belongs to the rare nemophila family (Hydrophyllaceae) and therefore adds to the variety of species grown on the farm. It inhibits nematodes and helps to improve soil health and structure.

An excellent way of combatting weeds is to sow a rapidly growing summer fodder mixture of field beans, peas, vetches, sunflowers, maize, vetchling and oats. Cultivate twice yearly and use for soilage or silage.

Row crops are often effectively preceded by a weed-suppressing and structure-improving catch crop (Landsberg mixture – fodder vetch, crimson clover and Italian ryegrass; Italian ryegrass and crimson clover; vetch and rye; turnip rape).

In southern Germany, and especially in damp areas, slender foxtail or black grass *(Alopecurus myosuroides)* is a troublesome weed grass, difficult to control chemically and causing serious reductions in yield if it gains the upper hand. The seeds ripen and drop during the summer weeks and germinate mainly at the end of September, i.e. at the same time as winter cereals. Winter barley is the first to be affected, but dense and vigorous growth permits it to stand up to the competition. Winter wheat, on the other hand, needs a long time to emerge in autumn and will only tiller by the end of winter, and in bad weather conditions not until early spring. Slender foxtail (black grass) can make good use of such an opportunity, so that a field of winter wheat may look very much like a meadow in spring. Hoeing will only have a limited effect and the best method is to drill the wheat late between the end of October and about 10 November – having first used tillage implements to combat the weed. 'Modern' combine harvesters unfortunately no longer allow one to collect the weed seeds at harvest time.

Mechanical measures

Modern technology (combine harvesting) has made harvesting dates much later. The result is that weeds reach full maturity and scatter their seeds. Table 24 lists the major species involved.

Combine harvesters also blow the chaff all over the stubble field, so that post harvest tillage has become most important. Stubble ploughs and grubbers cause weed seeds to germinate and help the following catch crop to grow apace, suppressing undesirables. Sufficient rain and adequate applications of dung liquor will strengthen the crop and enhance the effect.

During dry autumns, troublesome perennial weeds (e.g. couch grass) can be effectively controlled by repeated tillage operations alternating deep loosening with harrowing, so that the rhizomes dry up. No catch crop will be grown in this case. Root sections that remain in the soil use up their reserves and finally perish from exhaustion in late autumn if deeply buried. It is evident, then, that a plough can be effectively used in weed control by combining it with deep loosening and surface tillage. This applies particularly to spear thistles with their active growth in spring and seed production in summer, so that they are exhausted by early autumn. Their stolons thrive mainly on ploughpan which needs to be broken up and if possible resolved completely. Complete success will however depend on the farmer's personal engagement and observation. A spade test done year after year in critical sites is highly commended.

8. POST DRILLING CULTIVATIONS

weed	% of total seeds released
wind bent grass *(Apera spica-venti)*	95–100
slender foxtail *(Alopecurus myosuroides)*	65–75
wild oats *(Avena fatua)*	65–95
ivy-leaved speedwell *(Veronica hederifolia)*	100
common field speedwell *(Veronica persica)**	60–95
common chickweed *(Stellaria media)*	60–90
pennycress *(Thlaspi arvense)**	40–70
shepherd's purse *(Capsella bursa-pastoris)*	40–70
charlock *(Sinapis arvensis)*	55
hairy tare *(Vicia hirsuta)*	55
smooth tare *(Vicia tetrasperma)*	55
cleavers *(Galium aparine)*	20–40
black bindweed *(Fallopia convolvulus)*	20
redshank *(Polygonum persicaria)**	occasional
pale persicaria *(Polygonum lapathifolium)**	at most
wild camomile or scentless mayweed *(Chamomilla recutita)*	low

* Low-growing species; retained seed not picked up by combine.

Table 24. Percentage of seed released by weeds before winter wheat and spring barley are harvested with a combine (Koch & Hurle 1978).

A good autumn furrow will not only bring thistle and couch grass stolons to the surface but also those of coltsfoot and other weeds. This interrupts vegetative reproduction, the parts desiccate and decay on the surface or are attacked by frost. A spring furrow will not give the same result; instead, it brings seeds to the surface that germinate quickly and establish a new weed population.

The combination of deep and superficial tillage in alternation, and the sowing of a catch crop will also deal with curled dock. Renius (1978) advises drilling a ley and after the second cut ploughing as deep as possible, using a coulter, so that the dock is buried deeply. A vigorous, well-manured catch crop will prevent new shoots from coming up. The method may be used after winter barley that can be cleared early. Skimming at the end of July or in August (to cut off the roots below the collar) is followed by three to five deep grubbing or loosening operations at weekly intervals, after which winter rape or Landsberg mixture is sown.

In spring, the float or harrow is always used before putting in cereals or roots. This will cause the first generation of associated weeds to germinate, with the seedlings destroyed in the drilling process. The harrow should be used again three or four days after drilling, followed by further operations, preferably using a weed harrow, before the first tips appear and then again at the four leaf stage. A weed harrow is more flexible and adapts better to uneven ground; it has greater effective width. The speed is about 12 km/h (8 mph), compared to 6 km/h (4 mph) with a harrow. Both bury weed seedlings.

Mechanical hoes are used on a routine basis on row crops or vegetables – e.g. potatoes, beet, maize and carrots. They are now also being used again on cereals, the advantage being that times are less critical. Mechanical hoes can be used until the plants close up or shoot development starts; it is contraindicated only if wheel tracks are clearly causing damage in rainy weather or when there is a risk of frost.

The following points are to be considered when working with mechanical hoes:

- Both annual and perennial weeds are attacked.
- The method can be used for a relatively long part of the period of active growth. Early winter crops in good sites will tolerate a first hoeing operation in late autumn, which is then easily repeated in spring.
- In autumn, the share should not go below 0.5–1 cm (¼–½ in), whereas 3–4 cm (1½ in) is advisable in spring.
- Spring hoeing will stimulate further weeds into germination, so that winter crops may be contaminated with typical summer crop weeds and grasses unless further hoeing operations are done. These should be shallow to avoid damage to superficial roots.
- There is also an earthing-up effect. This does not really matter with cereals but can be a disadvantage for a

8. POST DRILLING CULTIVATIONS

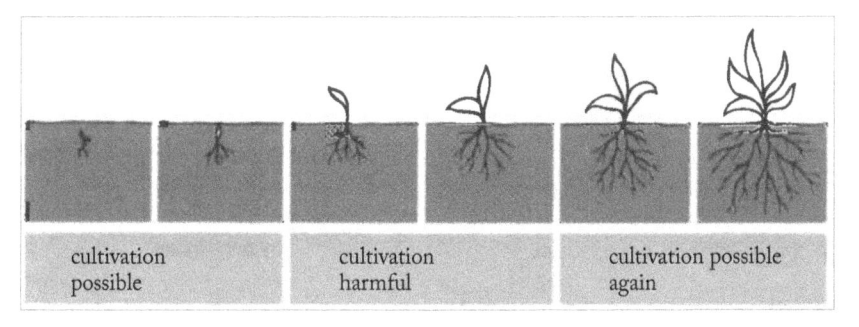

Figure 17. Growth stages of cereals when harrow and weed harrow may be used (Koch & Hurle 1978).

following ley, as uneven ground makes mowing more difficult.

- If the share is 10–12 cm (4–5 in) wide, the distance between cereal rows should be about 18–20 cm (7–8 in).
- With an effective width of 2.5 metres (8 ft) the time required to hoe one hectare (2½ acres) is about 3 hours for the first pass. About 2 hours should be reckoned for subsequent passes.
- Hoeing also has a positive effect on soil structure, it stimulates mineralization and reduces evaporation.

These benefits of mechanical hoeing should not blind us to the questions that remain open, nor indeed to its disadvantages:

- There are no problems with row crops, and if weather conditions are good both winter and spring wheat and spring barley come to no harm.
- Extreme caution is needed with oats and rye because of their shallow roots. In both cases the decision will depend on the state of the crop, the soil and the weather. There should be no more cold nights to follow, as sensitivity to frost will be heightened. Winter barley should not be hoed at all.
- Hoeing sometimes causes cereals, especially oats, to ripen unevenly, which is due to increased tillering.
- In sites where the climate is poor and very harsh, wide

	no. of weeds prior to cultivation		percentage of weeds after harrowing		
growth style	absolute	relative	unharmed	torn out	buried
cotyledons	2834	100	44	5	52
small rosette	5772	100	64	7	29
large rosette	5627	100	82	5	13
mean		100	64	6	31

Figure 18. Effect of harrowing on total weed populations at different growth stages (Koch & Hurle 1978).

distances between rows are bad for soil structure. Crops will be late closing up or never do so.

❦ In stony fields, stones will often damage seedlings or cover them up, so that yields are reduced.

Hoeing operations applied to cereals thus call for personal experience and careful observation.

The method should be more widely used in cereal cultivation, combining it with the use of weed harrow and harrow. Crops respond by growing at a better pace and weeds are kept down.

Thermal control

Organic farmers use steam and propane gas for thermal weed control. Steam sterilization is a completely nontoxic way of treating soil, compost heaps and greenhouses. Small appliances are handy for the purpose, or contractors may be employed using large units, mostly in horticulture.

Flame weeding with propane gas is coming to be more and more widely used on farms, especially on field-grown vegetables. In biodynamic agriculture, the first trials were made in 1967. In suitable weather conditions and with some basic knowledge, the implements can be used to good effect. The method is nontoxic as the combustion

8. POST DRILLING CULTIVATIONS

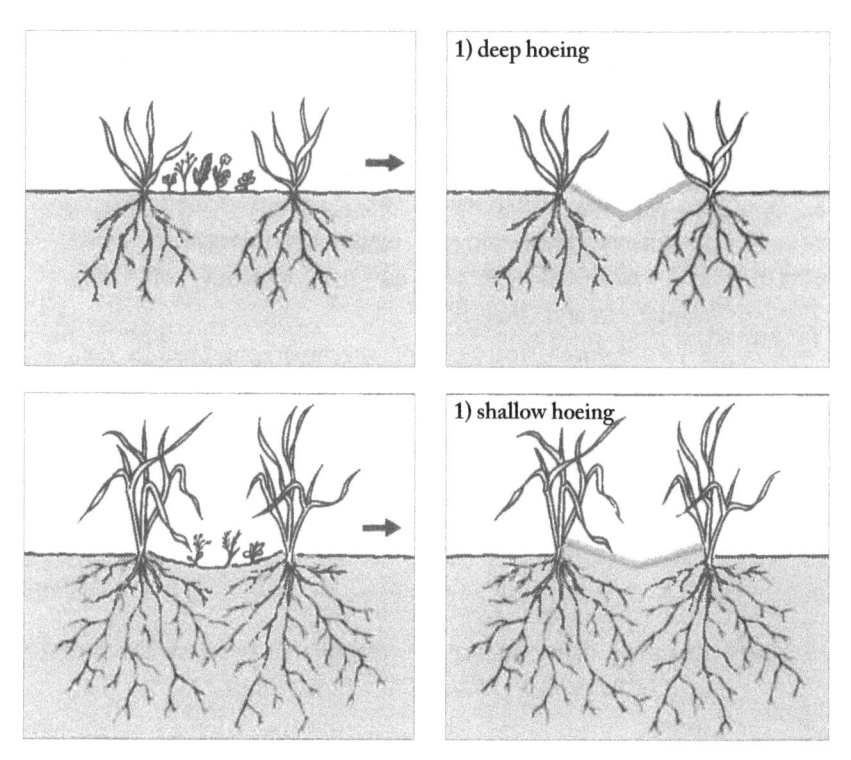

Figure 19. Effective use of mechanical hoe with cereal crops.

products of the gas are carbon dioxide and water. The weeds are not actually burnt. The rapid rise in temperature causes the fluid in the cells to expand and burst the cell walls. Death is also due to the fact that plant proteins coagulate at 50–60°C (120–140°F).

At first the only visible effect is that the plants turn a darker green. Pressure marks left by one's fingers demonstrate that damage has occurred.

A flame weeding operation done at the right time will destroy numerous seed-bearing and grass weeds. Weeds with vigorous root systems such as thistles and dock do resprout, however, so that the operation may have to be repeated a number of times.

Flame weeding causes minimal if any damage to soil organisms, for at a speed of 1–7 km/h (1–4 mph) the high temperatures do not penetrate the soil to any depth.

M. Hoffmann (1980) distinguishes four methods:
1. *Total:* Burns all plant growth over the treated surface area. To be regarded as a preventive measure only.
2. *Selective:* Used to remove undesirable weeds in crops that are relatively heat-resistant (woody stems). Good results have been noted and the method is increasingly used with maize crops, in tree nurseries, orchards, soft fruit plantations and vineyards.
3. *Pre-emergence:* Treatment of vegetable rows before the plants emerge (use of pressure wheel). Hand roguing may follow later.
4. *Spot:* Deals with weeds that form colonies and persistent weeds. The method also gives good results in gardens, farmyards, around shrubs and along roadsides.

Different types of appliances may be used:
- Knapsack flame weeder for very small areas; may also be used in fields that cannot take vehicular traffic in very wet springs.
- Hand-operated two-wheeler flame weeder; useful in vegetable fields and when larger vehicles cannot be used after rain.
- Large add-on units for use with tractors; a number of designs are on the market.

Pre-emergence flame weeding effectively controls early weed growth in vegetable crops that are slow to germinate, such as carrots.

A practical example from a farm

It is usually only older farmers who are still familiar with mechanical weed control. Over the page, Guenther Count Finkenstein describes the methods used on his farm.

8. POST DRILLING CULTIVATIONS

Winter cereals

To permit the use of hoeing machines – this appears necessary on all heavy soils – we drill with spacings of 17–20 cm (7–8 in).

For good results the duck's foot should be at least 10 cm (4 in) wide. The difference between row spacings and width of shares should be not less than 7 cm (3 in) to allow the implement to be guided. Depending on the effective width of the seed drill, it is a good idea to block outlets matching the wheel distance of the tractor used for post drilling cultivation; this reduces damage from cultivation operations. About 40 cm (16 in) per wheel is usually sufficient. Losses due to this are within reasonable limits for a 6 metre (20 ft) drill. With narrow drills, one way of keeping losses down would be to position the gaps as for the light break method, and another would be to use narrower wheels.

Drilling needs to be done with reasonable care. The first post drilling cultivation using a weed harrow on winter barley and rye may be done just before or at first emergence, but we have no experience of this. We hardly ever grow winter barley and with rye there appears to be little need, especially when there is so much other work do be done in autumn. It is highly likely, however, that the method will be successful in controlling some difficult weeds. The soil must be firm enough so that the harrow does not tear out the germinating grain that is lying just below the surface.

Rye crops with a serious weed problem need shallow hoeing.

The first spring cultivation depends on the condition of both soil and crop.

If frost has lifted the soil, a Cambridge roll is used when the soil is dry enough. The plants must be of a size where not too many will be buried as clods are pressed flat. After rolling it is usually necessary to wait a day, as the crop has been flattened. It must be upright again before the weed harrow is used, otherwise the risk of plants being buried is too high.

If the soil has puddled after a wet mild winter, rolling would cause serious damage; what the soil needs in this case is to be torn open.

Like any other operation used on weeds in the germinating or cotyledon stage, early use of the weed harrow is much more effective than anything that is done later. A second pass may be made at any time prior to hoeing.

Once the plants are big enough to cope with the slight ridging effect, a mechanical hoe with 10–12 cm (4–5 in) wide shares is used, as already mentioned. When a rye crop is badly fouled with black grass or foxtail (rye is more sensitive to this than wheat) we use 12 cm (5 in) shares and work at a slow pace. We use tool carriers and centre-mounted frames, with the harrow connected to the rear hydraulic element. This has proved particularly effective in wet years, with weeds growing between rows either cut into and ripped out or chopped off and dragged away. This reduces the risk of weeds taking root again after the next rain.

A question that keeps coming up is whether mechanical hoes can be used on rye. According to reliable reports from W. Renius, accurate trials run after the First World War showed no increase or loss in yields. Appreciably fouled rye therefore needs shallow hoeing, particularly if the weed in question is foxtail.

During the first years after conversion our winter wheat was so badly fouled with foxtail that one could no longer see the rows in some parts. One pass with the hoeing machine followed by a pass with the weed harrow gave the crop sufficient air. The mature crop was reasonably weed free, despite the fact that duck's foot harrows will only cut about 50% of the weeds. Wheat offers powerful competition to foxtail growing in the rows. An essential precondition is sufficient crop vigour and proper management. Initially we had been rather doubtful about the method, but thanks to these encouraging results the appearance of foxtail in cereal and other crops no longer constitutes a major problem.

The good thing about mechanical weed control is that specific intervention in the weed population allows us to retain a desirable variety of plants at a level where they do not compete with crops. Russian researches into phytoncides suggest that certain arable weeds have direct or indirect positive effects on crops.

Spring cereals

Except in extreme cases, weed control can be entirely mechanical with spring cereals. It does need good drilling technique under dry conditions and the kind of good crumb structure you get with an autumn furrow or on soils that can be ploughed in spring.

We normally drill at a depth of not less than 2 cm (1 in); otherwise the crop is not sufficiently anchored and may be ripped out during the first pass with a weed harrow. Rolling after drilling prevents the harrow from going too deep, especially in the first and second passes. If the soil is still rather fresh (moist) after drilling, we wait a few days before harrowing. Rolling should however be done before the cereal has produced the first fine rootlets. From then until the white leaf shoots pierce the surface, vehicles should only go on the field if tramlines can be followed exactly. At this stage of plant development, any new tracks or hoof marks from draught animals will cause gaps.

If rain firms the soil after drilling, rolling will not be necessary. It is still possible to roll when the crop is just emerging, but the timing is critical, so that one needs to be ready for instant action at this time. The weed harrow on its own may also be used between light showers or if necessary on a Saturday afternoon or a Sunday at approximately 5 ha/hour (12 acres/hour).

The first harrowing pass should be made at first emergence at the latest. Once the cotyledon has opened out, no cultivation can be done until the 3 or 4 leaf stage, as the tender seedlings will die if covered by small clods of earth or by stones.

Weeds not destroyed at the time of first emergence of cereals will be in the small rosette stage when the cereal crop has developed 3 or 4 leaves and can then only be controlled to a limited extent. In cases of serious fouling this may result in failure. It is therefore vital to get the work done at the right time, another reason being that this will reduce the time during which weed control is impossible to its minimum. Potential weather problems are reduced by blind harrowing a few days earlier, when the white leaf spears in the soil have not yet come to the point of emergence. Blind harrowing will in any case be required as an additional measure if the weed problem is severe.

Timing is vital with mechanical cultivation operations.

Early passes with the weed harrow can be done rapidly and should be shallow. From the 3 or 4 leaf stage onwards passes must be made slowly to avoid covering the seedlings with small clods of earth or stones. One has to get down from the tractor, take a look and adjust the speed as required.

If the right times for blind harrowing and first emergence work have been missed, duck's foot shares need to be brought into play as soon as possible. This has to be done at low speed to avoid burying the crop in the ridges. After this, the method is the same as for winter cereals.

We take account of the positions of the stars whenever possible, basing ourselves on Maria Thun's calendar. For work that has to be done within a certain time, e.g. for using the weed harrow at first emergence, we merely aim to avoid unfavourable times. In planning the week's work, we try to coordinate other necessary cultivations according to their favourable times. If the weather, the labour situation or other reasons make this impossible, we limit ourselves to avoiding unfavourable periods.

9. Arable Crops

Cereals

As mentioned in the historical review (p. 44), most cultivated plants originate from a few limited areas on the globe, mainly the uplands of subtropical regions which are known as sites for the gene pools or centres of variation. As may be seen from Figure 20, the seven cereals – wheat and barley; rice or millet; oats, rye and maize – come from the different continents (except for Australia). The centre of post-Atlantean civilization, the Iranian uplands, was the original home of wheat and barley, which are probably the oldest grain crops, and of many other crops such as lentil and pea, flax, mustard, vetches and numerous fruit and vegetable species.

In Central Europe, wheat is the principal bread grain today, with rye coming second; malting barley is specifically grown for beer brewing, and the other cereals serve as feed-stuffs in animal husbandry and intensive livestock farming. Almost half the world population lives on rice, whereas millet is much less eaten now than formerly.

Almost two thirds of the arable land in the western part of Germany is used for cereal growing, with cereals advisedly making up between 30 and 60% of the cropping sequence. Depending on local conditions, a good balance can be achieved between legumes and oil crops on the one hand and cereals and row crops on the other, providing for harmonious alternation of phases during which humus builds up and is depleted, a good crumb structure is established and reduced and nitrogen is fixed and utilized.

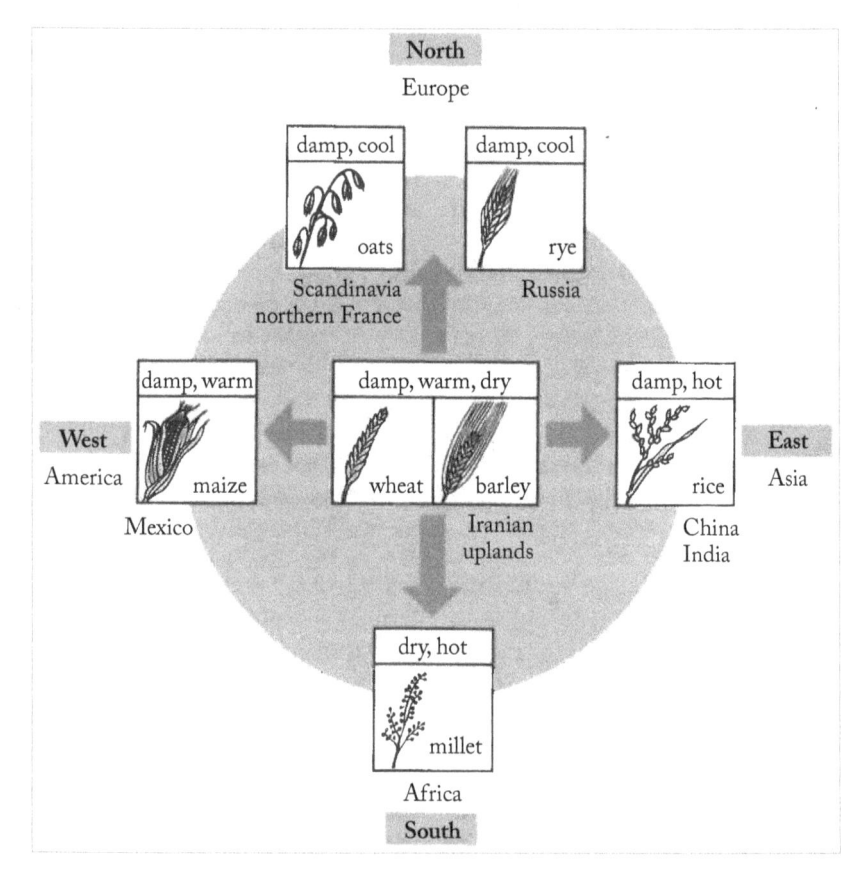

Figure 20. Origin of the seven cereals.

Cereal growing is important from the point of view of labour intensity. The work is easily mechanized, and on a highly specialized farm it takes about ten hours of labour per hectare to do everything that is required.

Cereals belong to the family of grasses (Gramineae), with winter and spring varieties. Many different varieties offer a wide range of characteristics, so that it is easy to choose one that meets local needs regarding drilling time, weather, soil and fertilizer requirements, intended purpose (e.g. wheat for bread or pasta production) and harvesting time.

9. ARABLE CROPS

All cereals require a firm seed bed but winter crops especially so, for plants need to be firmly rooted to survive both intermittent and black frosts. Winter barley and rye need to be drilled at low depths and in good time, as tillering should occur in autumn. Proud growth due to unusually mild weather conditions can be corrected by letting sheep graze under careful control. If this is not done and there is extended snow cover during winter, the crop may suffer damage from snow mould.

In a mild climate, cereals may be hoed in autumn, with the slight ridging effect providing added protection. If drilling is done later, so that hoeing will not be possible, drag rings fitted to the seed share on light soils or pressure wheels and coverers on heavier soils produce the same effect, with the seedlings growing in a slight furrow, protected from wind and black frost by the walls.

Winter damage may also be prevented by the following methods:

- Use a harrow or roll to break up snow crusts and prevent lush growth from asphyxiating in frost-free ground.
- Roll frost-lifted soil in early spring so that rootlets that have become exposed are again brought into contact with moist earth.
- Use a harrow and roll when active growth is in progress to break up puddled and crusted soil.

If conditions permit, a thin layer of fully humified compost applied to the field will give added protection and cause the soil to warm up more quickly in spring. Further cultivation with weed harrow, harrow and hoeing machine is as described above.

The ear or panicle begins to develop early, at the time when the plant tillers, and this is the time for a first application of horn silica. Tillering may start in autumn with winter barley and rye and frequently also with spelt and winter wheat if drilled early. An autumn application of horn silica has therefore been found to be of definite benefit under favourable weather conditions in mild climates. Results have been less good in harsher climates.

The addition of a drill fertilizer to winter and spring cereal seed has proved effective. The following may be used:

1. Drill Oscorna (a mixture of horn, bone, blood, meat and feather meal, with large particles removed by sieving).
2. 60% Drill Oscorna, 20% basalt rock dust, 10% calcified seaweed.
3. 25% Drill Oscorna, 10% basalt rock dust, 10% calcified seaweed, 55% fine humified compost (maximum moisture level 20%).

Nos. 2 and 3 really need a two-hopper drill to work efficiently. Add 10 kg or at most 20 kg of the chosen mixture per 100 kg of seed and mix thoroughly. The mixture may separate, and it is therefore advisable to put in only small quantities at a time and check that there is an even flow of the right number of seeds. Drill fertilization will usually
- accelerate and improve root development,
- encourage tillering in thinly sown crops,
- promote ear and grain development and
- frequently result in earlier maturity and harvest.

On sites where conditions are otherwise not exactly ideal (e.g. unfavourable annual weather conditions) wheat will have a higher gluten content and quality.

Wheat

Wheat *(Triticum aestivum)* is the most exacting cereal crop; being less able to release and utilize minerals, it needs soil with a good supply of nutrients and lime. Wheat does best after preceding crops like rape, legumes (peas, field beans, clover and clover/grass ley) and row crops that can be cleared early.

The seed bed needs to be firm and not too fine on the surface. Wheat is the cereal that tolerates late drilling best of all.

Although the most resistant of all cereals to mechanical cultivation with weed harrow, harrow and hoe, it tends to react rather sensitively to massive competition from weeds. In the west of Germany, winter wheat has the greatest number of varieties (*c.* 50), with winter barley coming second. It is not as adaptable as rye (*c.* 10 varieties), but it is always possible to find the right variety for the intended purpose and the given situation. Most spring wheat varieties have good blending properties and may be added to less good winter wheat to improve baking quality and achieve a mean moist gluten content of *c.* 30% at a Q_0 of 16–22.* Coarse meal from grain of this quality will produce well-flavoured wholemeal bread.

Spring wheat needs to be drilled as early as possible from the end of January, depending on local conditions. Unlike winter wheat it does not tolerate being 'rubbed into' the soil. Regarding preceding crops its needs are the same as those of winter wheat. Clover and grass/clover leys are best, especially as they allow a late cut for soilage in autumn. In areas where there is a risk of winter kill, it is advisable to grow spring wheat on part of the wheat area; it will give better yields than winter wheat that has suffered damage. Fields occupied by winter cereals also tend to have more of a weed problem. Spelt (*Triticum spelta*) is probably an ancestor of our ordinary wheat (*Triticum aestivum*) and is only grown to a limited extent. There has been a slight increase in demand for dried unripe spelt and special baked goods. The glumes remain attached after threshing, so that spelt needs an additional tanning process to remove them. The glumes are left on for seed. The grain is flinty and the flour a pale yellow.

Spelt is less exacting where soil and climate are concerned, except that it needs slightly more lime than wheat does. Its preceding crop value lies between that of wheat, which is less, and rye.

Rye

Rye *(Secale cereale)* is the most frost-resistant of the cereals; it is classed as a secondary crop plant, which means that it was initially a wild plant found in wheat and barley fields and only gradually

---

* Q_0 is swelling number, a unit used to express gluten quality. For winter wheat, a Q_0 of more than 16 and a moist gluten level of more than 22% are desirable.

developed into a plant of economic value. It benefited from its ability to inhibit the growth of other plants – both weeds and crop plants – and suppress them, probably by excreting growth inhibitors (allelopathy) in the root region. Rye has three seedling roots and rapidly develops a root system that spreads horizontally at first and later also downwards. Contrary to common belief, the nutrient requirement for an equal yield is higher than for wheat. On the other hand the root mass and root performance are greater, so that adequate yields may be achieved even in relatively poor soils and climates, including podsolised heath and woodland soils and lean shallow slate-based soils.

Rye needs a fine, firm seed bed to ensure shallow, even drilling and prevent winter kill due to frost lifting. The soil needs to be dry for drilling. Careful use of weed harrow and harrow is well tolerated in spring (caution is needed if there is a threat of frost). Hoeing is indicated only if there is a serious weed problem.

In spring, growth will be rapid once the soil has warmed up, so that rye also makes a good fodder catch crop, either on its own or mixed with winter vetches. It is resistant to black frost, being able to survive temperatures down to −30°C (−20°F). It is less resistant to intermittent frost, waterlogging and high snow levels, as there is a risk of rot and extensive snow mould (a fungus). Rye is relatively sensitive to cold at the flowering stage. Late frost may be responsible for uneven ears, so that it is usually safer to grow spring rye on sites where there is a risk. Spring rye is not much grown. It may do better than winter rye on very light sandy soils or in areas where waterlogging is common in winter; otherwise it is only grown, if at all, to make up for winter kill. Drilling should be as early as possible.

Schmidt rye has been used to some extent in biodynamic trials.* Rye is particularly difficult to breed as it is a cross-pollinating crop, self-sterile and sensitive to inbreeding. To maintain a variety and its yield for any length of time, care must be taken to keep a good distance between different varieties of rye; the distance should be

---

* Schmidt rye was bred by Martin Schmidt in Germany, using his own specially developed method (ear plot selection).

600 metres (2000 ft), and 1000 metres (3300 ft) in the main wind direction. Otherwise alien pollination may result in hybridization.

Ergot is best prevented by mowing baulks early, so that the fungus does not spread to the rye from grasses that flower earlier.

Barley

Barley *(Hordeum vulgare)* is one of the oldest cultivated plants if not *the* oldest; it has played a major role as a food in the past, mainly in form of gruel and flatbread. Today barley is grown mainly for brewing and fodder, almost a quarter of the arable land in western Germany being used for the purpose.

Winter barley is generally used for fodder. In recent years, two-rowed winter varieties have been increasingly grown for malting. It requires timely drilling in firm soil having a good lime status. Winter rape is the best preceding crop, with vegetable peas, early potatoes and early-clearing grain peas to follow. Barley is an exacting crop where weed control is concerned, but does tolerate careful harrowing and hoeing. Its early harvesting date makes it a good nurse crop, providing grain fodder in early July.

> The earliest finds of barley as a cultivated plant in the Nile Delta go back to predynastic times, those in northern Syria and Assyria to 4000–4500 BC.

Demand for human consumption is limited, and spring barley is best for the purpose, especially glumeless varieties. Barley grown as a cereal for special diets should have similar qualities to malting barley: glumed barley should have 8–12% of fine glumes, 75–85% yield of malt, a germinating power of not less than 95%, good smell and colour, high thousand grain weight and low protein content (less than 12%). These characteristics are enhanced by early drilling in a fine crumb. Row crops, above all potatoes, have proved to be the best preceding crop; they leave the soil weed-free and well-structured, full of life but not liable to force growth. Any kind of fertilization tending to force growth must be avoided as it causes increased protein levels that are liable to cause severe flatus. The seed rate needs to be fairly high, otherwise increased tillering will result in uneven growth and maturity. Whilst barley makes a good nurse crop for clover, quality may be lost and harvesting made difficult if the clover growth is too vigorous. The clover therefore needs to be drilled at a later date, e.g. after tillering.

Cattle like eating barley straw with dried undercrop hay.

Oats

Oats *(Avena sativa)* developed relatively late as a crop but then spread rapidly in the damp and cool parts of Central Europe, where they were the staple food for a long time. The protein content is one of the highest among cereals and the fat content the highest. Oats have a high nutritional value because of their high iron and calcium levels. Oat gruel and porridge are important for children and the infirm. Young stock also do well on oats. In temperate regions oats are fed to horses, taking the place of the barley that is fed in hotter regions (e.g. Arabia).

Oats have the most effective root system, with good yields even from a second or third successive crop. Owing to a high capacity for dissolving minerals and their deep-rooting qualities, oats also do well on ploughed up long-term grassland.

Oats are considered the best preceding crop among the four cereals. They act as a cleaning crop in close cropping sequences, suppressing take-all.

Oats are an unexacting crop; high yields can be achieved on good sites and the best yields of all cereals even in unfavourable conditions. They do however require plenty of moisture. Dry conditions after stem elongation may seriously affect the yield. Winter oats are grown only on sites that are consistently dry in spring.

Naked oats demand better soil quality and are particularly sensitive to prolonged dry periods. They will then give low yields, compared to the 4 t/ha (1¾ tons/acre) that are not uncommon for this crop even on poor sites if weather conditions have been favourable. Naked oats are in demand for human consumption and also make excellent feed for young stock, but it is advisable to grow them only in areas where the climate suits. Experience has shown that they tend to revert to glumed oats, so that new seed has to be bought in every two or three years. Storage of the harvested grain requires great care as it tends to pack down tightly and easily grows mouldy, decreases in germinating power and acquires a musty smell and taste.

Oat straw harvested dry and in good condition provides excellent fodder for young stock and supplementary feed for cows because of its relatively high phosphorus, potassium and magnesium content. Early maturing oat varieties make a good nurse crop for oversown crops, with the longer straw preventing the undercrop from growing through, a risk that exists with spring barley. In dry areas or years, competition for soil water from the undercrop may reduce yields.

Cereal cyst nematode will attack spring wheat and barley and a number of grasses as well as oats and may cause serious losses in cropping sequences with a high proportion of cereals. The only way out is to cut down on oats in favour of winter crops and to intensify the use of catch crops.

Smut and seed treatments

The exclusion of chemical seed treatments may result in an epidemic spread of a number of fungus diseases, the most important of these being covered smut or bunt *(Tilletia caries)*, which may cause up to 50% losses. Transmitted via the grain, the fungus inhibits reproduction.

> If seed is not treated, various fungal diseases may cause serious damage.

The warm water treatment, originally introduced to combat loose smut *(Ustilago nuda)* in wheat and barley, effectively killed the fungus for many years but failed to do so on many farms in 1984. This suggests that resistant strains have developed. The Biodynamic Research Institute in Darmstadt, Germany, is currently conducting trials with physical and biological methods to combat covered smut. The fungus is transmitted by spores adhering to the grain and safe ways have to be found to remove, inactivate or kill these.

A grain washing plant will remove up to 95% of the spores, depending on intensity. A simple practical method is to wash the seed in a concrete mixer. Subsequent treatment with cattle dung liquor, calcified seaweed and wood ash or fungicidal plant extracts (e.g. mustard or allium oil) will enhance the protective effect. Seed low in vitality is known to have low resistance to fungus attack, and it is advisable to do everything possible to improve seed quality and hence also natural resistance. This is all the more important in view of the fact that infection with covered smut when seedlings are more than 2 cm (1 in) in length will merely result in latent disease. Seed baths with extracts of medicinal plants (see page 79) and the use of horsetail tea or liquor as recommended by Rudolf Steiner are helpful in this respect.

The warm water treatment continues to be indispensable in dealing with loose smut. The method consists in immersing the seed in water at a temperature of 47°C (117°F) for two hours, using a thermostat, and then drying it in a cabinet at a temperature not exceeding 33°C (91°F) to a moisture level of not less than 15%, or better 16%. This will prevent damage to the germinating power, especially if drilling does not follow immediately.

Row crops

Hoeing was the earliest form of cultivation used when people first began to settle. Using digging sticks and primitive hoes, virgin soil was made sufficiently fertile to grow cultivated plants – mainly palms, bananas, melons, pumpkins, cucumbers and two kinds of tubers: taro and yams. Fields had to be abandoned after a few years and new ones cultivated,* as yields would be getting low. Even today millions of people base their existence on this type of agriculture, most of them in the tropical regions of Africa, Asia and South America.

Sheep and goats were the first domestic animals, probably soon followed by pigs.

These developments generally went hand in hand with a matriarchal society, where women did most of the hoeing in the fields. The development of patriarchal societies brought the domestication and breeding of horses and cattle. Ploughing then became possible, and cereals came to be bred and improved as the great early civilizations evolved.

Row crops, or 'hoed crops' as they are called in German, grown in temperate latitudes today are potatoes, beet, beetroot, carrot, cabbage, maize for silage and the majority of field vegetables. All give high yields per area, and intensive cultivation means a reduction in weed populations and enhanced mineralization of the nitrogen stores collected by a high volume of leguminous crops. Farmyard manure improves yields even further. Manuring and labour requirements are relatively high. These plants consume much humus; 15–35% of them in a cropping sequence will on average deplete dry organic matter by 2–4 t/ha (1–2 tons/acre). Labour is required during different periods than for cereals and, providing the proportion is not excessive, row crops help to maintain a balanced work programme.

In biodynamic agriculture in Germany, the most important row crops are:
- potatoes grown for food
- beetroot and carrots for juice and for food
- cabbage for making sauerkraut
- mangels and swedes for fodder, and
- field vegetables for immediate sale or winter storage.

---

*'Cultivate' derives from the Latin *colere, cultus*, to inhabit, protect, honour with worship.

Maize for silage is only grown to a limited extent, e.g. to bridge gaps in fodder supply in August and September in a dry year.

Potatoes

The potato *(Solanum tuberosum)* comes from South America, where it has been grown in the High Andes since 3000 BC if not before. The first tubers reached Spain in about AD 1560, but it was not until the eighteenth century that this crop with its high area yield became the staple food of Europe. Good harvests initially helped with the hunger years that occurred at regular intervals. More and more potatoes were grown, but then in the 1840s late blight *(Phytophthora infestans)* caused potato famine all over Europe. Crop failures due to this were the reason for massive emigration from Ireland to America from 1845 onwards, with a vast reduction in the Irish population.

Before potatoes were introduced, cereals and pulses were the staple diet in Central Europe. It is known that special eating habits have a marked effect on inner development in humans ('you are what your eat'). A diet with a high proportion of meat tends to encourage aggressive tendencies but also heroism and courage. A vegetarian diet usually results in more gentle behaviour but may also lead to fanatical propaganda for specific life styles. Potatoes as a stem element growing in the root region address mainly the nerves and senses, especially the midbrain. An almost exclusive diet of potatoes prepares the ground for materialistic ideas and actions (Steiner, 1923, lecture of Sep 22).

The potato is a member of the nightshade family (Solanaceae) and its reactions to light are peculiar. On exposure to sunlight the green parts of the plant produce the toxic alkaloid solanin. The nontoxic 'fruits', which are tubers and have not arisen through a flowering process, ripen in the dark earth to provide food for humanity. If precipitations of unusual force or cracks in the soil caused by dryness expose these to the light, they too will turn green and become toxic.

Unlike plants that have more of the nature of light in them, above all the cereals, potatoes produce lush vegetative growth, usually a very dark green, of the kind normally only seen in plants that grow in the shade.

They respond quite definitely to intense use of the biodynamic field sprays, especially horn silica, the light mediator.

Soil and climatic requirements

Potatoes are highly adaptable as regards both soil and climate.

Extremely heavy, wet and cold soil conditions are least suitable, whilst loosely structured soils rich in humus that allow air, water and warmth to move as required will encourage growth, health and good yields. High pH values (sandy soils above 6.0, good loams above 6.5–6.9) mean increased risk of common potato scab (*Streptomyces scabies*), and liming will increase the tendency.

The haulm is sensitive to low temperatures – it may suffer frost damage even at −1.5°C (29.5°F) – so that sites exposed to frost are not really suitable. If late frosts kill the young shoots at a very early stage of development, new shoots will usually grow from dormant eyes and yields may still be good if weather conditions are favourable. It is however not advisable to use plants that have been subject to late frost for seed, for such weakened plant material is more susceptible to degenerative changes. This is also why cultivations carry more of a risk in such sites. Early frosts in autumn on the other hand do not have such negative consequences. Tubers may still make good growth in the milder weather that follows, even if the leaf mass has been much reduced.

Potatoes do well in areas where average daily temperatures are in the region of the 20°C (68°F) isotherm. The water requirement only increases at the flowering stage, when the plants need a good, reliable, but not excessive supply.

Preceding crops

Suitable preceding crops are those that leave a well-structured, friable soil with a good root space, above all legumes, and on sandy soils mainly lupins and serradella. When stubble-sown after field crops that clear early, e.g. winter rape, winter barley and winter rye, lupins will show vigorous growth by autumn; they may be used for fodder or, better, allowed to remain over winter, exposed to frost killing and then ploughed in in early spring. This will give ideal conditions for a good potato harvest. Green manuring also helps to reduce the risk of scab.

More or less the same holds true for Landsberg mixture (see page 190) in areas where a spring furrow is possible. Ploughed-in permanent grassland may also be considered, unless there is a problem with white grub or wireworm. On heavy arable soils, leys of several years make the best preceding crop. As with permanent grassland, ploughing needs to be done in good time in late summer; otherwise harvest residues and root residues hinder cultivation and nitrogen mineralization in spring.

Potatoes are largely self-compatible; the only possible problem with successively grown crops would be black scab or eelworm.

Soil preparation and manuring

Cultivations should aim to give the soil a loose structure. This ensures good growth and clean work with special planting and harvesting implements. Medium and light soils will take a spring furrow, others need to be ploughed in autumn.

For sound, well-flavoured potatoes with good keeping qualities, it is recommended to use 20–30 t/ha (9–13 tons/acre) of well rotted manure on the preceding catch crop. Experience at Talhof and other farms has shown that a further application of humified composted manure prior to planting helps plant development and reduces susceptibility to late blight and Colorado beetle. The addition of pigs' bristles guaranteed free of residues to the composted manure has a positive effect especially on heavy soils, whilst direct application favours fungus and Colorado beetle attack. If the quantity is too large, crop quality is reduced. The same applies to applications of dung or dung liquor liable to force growth, which also affect the flavour.

Seed and planting

Apart from mellow soil, a good preceding crop and appropriate manuring, seed quality is the major factor in growth and yield. Seed potatoes should

- come from well-matured crops,
- be the right shape for the variety,
- weigh between 35 and 60 g (1¼–2 oz), depending on variety,
- have been stored under suitable conditions, either in well made clamps or indoors at temperatures between +2° and +6°C (36–43°F); a relative humidity of 65–80% will improve keeping quality,
- not have been treated with sprout inhibitor,
- not have sprouted during winter storage, and
- be ready to sprout when planted.

Degeneration is more marked with potato than with any other agricultural crop. The term is used to define reduced yields due to increasing deterioration and increased virus susceptibility in home-grown seed. Stunted plants, gaps where seed has failed to sprout, diseases of the stem base and susceptibility to blight increase year by year and rapidly lead to crop failure.

> Degeneration is evident from reduced yields, increased susceptibility to pests and diseases and loss of germinative power.

To prevent this, conventional farmers use fresh certified seed annually in endangered areas or every two years on better sites.

The causes of degeneration are in the first place the many modern farming methods that are mainly designed to increase yield. On the other hand potatoes do adapt to a wide range of climatic and soil conditions and this is helped by choice of the right variety. Unsuitable environmental conditions will, however, weaken their vitality and constitution in the long run. It is also possible that breeding for resistance to black scab has increased susceptibility to viruses. This susceptibility originates from a single scab-resistant variety with which all other varieties were crossed in the 1930s.

The real cause is probably consistent omission of the sexual phase, i.e. growing the potatoes by vegetative propagation for generations.

The potato is an annual. The true seed produces remarkably delicate small plants with a single shoot and relatively vigorous root growth, developing tubers the size of hazelnuts. It needs two years of vegetative propagation to produce potato haulms and tubers of normal size. With vegetative propagation, on the other hand, the reserves of the parent tuber give rise to vigorous young plants with multiple shoots and a limited root system.

Healthy planting material can be obtained by growing potatoes from true seed.

Asked what could be done to improve the regenerative powers of potatoes, Rudolf Steiner told Ernst Stegemann to use a special chitting technique. The method is as follows. Mark plants in the field that show healthy, vigorous growth and harvest these separately (plant selection). Discard any tubers that are bad or show rot and store the rest in a separate clamp or place. Three or four weeks before the ideal planting date, the eyes are cut. Use only fully developed single eyes that are well apart from others and have not yet started to sprout. Cut off the heel and rose ends of the potato. Leave only the minimum of tissue around the eye, but enough to prevent it from drying out completely. Care must be taken not to damage the bud as this would increase susceptibility to blackleg *(Erwinia carotovora)* and stem canker *(Rhizoctonia solani)*. Wedge-shaped sections with the bud at the wider end are the best.

Mix with wood ash and fine siliceous sand and spread out flat in a well-lit place that is frost-free but not too warm. The buds need to come to the point where they are ready to sprout, producing 2–3 mm ($1/8$ in) long sprouts in the light. If planting has to be delayed, an application of horn silica made in the morning will inhibit further sprouting.

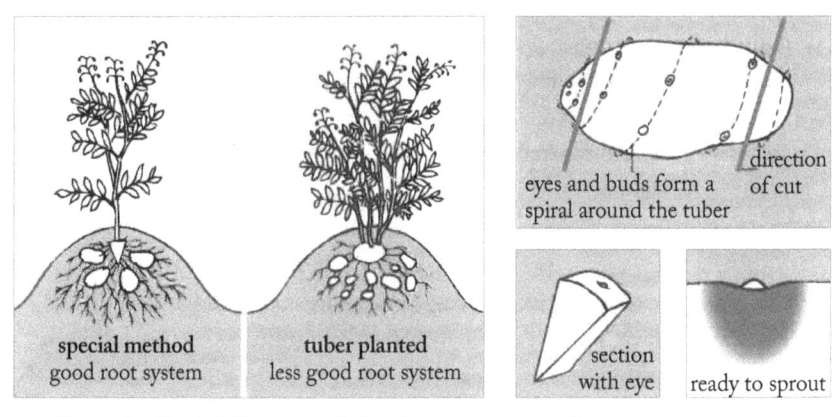

Figure 21. Special chitting method to improve potato seed.

The seed baths are given as follows: the first one about fourteen days, the second one about seven days and the third two days before the planned planting date. Valerian and oak bark have proved more effective with chitted potatoes than birch pit concentrate. Keeping the potato material moist for the week preceding planting will specifically stimulate root growth.

Make the rows with the planting machine and place the eye pieces by hand so that they lie flat, with the bud on top. Potatoes chitted by this method tend to produce large single tubers; reducing the distance in the row to 15 cm (6 in) counteracts this tendency, so that you get evenly medium-sized potatoes. The distance between rows is the usual 62.5 cm (25 in), or 75 cm (30 in) with larger tractors. With chitted material, development is slightly delayed compared with whole potatoes. In the right weather conditions the plants soon develop a vigorous root system, but herbaceous growth from the single compact shoot lags behind right into July. The active growth that follows generally produces vigorous, well-developed foliage so that the whole field looks a lighter green. Unless attacked by frost, the haulms persist longer in autumn. Ripeness is assessed by checking the firmness of the skin of the tubers.

During the period of active growth, two or three passes are made to remove all degenerated plants, putting these in a separate compost heap for use on permanent grassland only. Tubers from well-developed plants are selected to provide planting material for the next year. The rest are used to grow potatoes for food. Potatoes selected from $^1/_{100}$ ha and chitted by the above method provide material for planting *c.* $^1/_{10}$ ha. The yield from this provides the seed for 1 ha of table potatoes.

With more plants per hectare and the rose and heel ends cut off, the method does not give a saving in seed material, and it is also very labour intensive. Potatoes grown from seed produced by this method will sometimes have a more earthy taste.

The method is designed to provide healthy seed that is not subject to degeneration. It should not be used for any other purpose.

> The special chitting method is designed to produce healthy seed; it is labour intensive and may affect the flavour. The method is therefore not suitable for growing table potatoes.

Seed not set aside for special chitting is also taken from store about four weeks before planting and spread in a well-lit place that is frost-free but not too warm (*c.* 12–14°C, 54–57°F). Under these conditions the tubers shrivel and dormancy ends more quickly. The metabolic processes that precede sprouting – increasing conversion of sugar, cleavage of starch, degradation of inhibitors and rising concentration of growth hormones – are stimulated in the bud region, and compact sprouts develop in the light. The tubers are getting ready to sprout. As a result of this treatment, a satisfactory number of buds will sprout, something not even achieved with relatively low soil temperatures.

Seed baths, used as above, support these processes and also stimulate root growth.

The optimum number of plants/ha is considered to be 50,000 (20,000/acre). With rows 62.5 cm (25 in) apart, this means a distance of 30–32 cm (12–13 in) in the row. The distance needs to be shorter with smaller tubers grown for seed and longer if larger potatoes are to be grown for food. Depending on the weight per tuber, this gives a seed requirement of 1.75–3 t/ha (0.75–1.3 tons/acre).

Soil preparation should be such that weed control is on the whole complete 4 to 6 weeks before planting.

Planting depth is 5–10 cm (2–4 in), and should be as consistent as possible to avoid losses on clearing. A thin but complete cover of earth protects from frost and helps the soil to warm up. In heavy soils and with late planting, the tubers should be planted less deeply.

Cultivations

Post planting cultivations aim to maintain a loose, friable soil and build ridges that are as even as possible. Weed control needs to continue until the plants cover the ground in each row and are able to keep the soil friable and weeds at bay on their own.

Prior to emergence, alternating or combined use of ridging tool and weed harrow has proved effective; mechanical hoeing follows later. Cultivations need to get progressively more shallow as the crop develops, to avoid inhibition of growth through damage to the fine roots. A weed harrow (chain harrow) is sufficiently flexible to adapt to the ridges; working shallowly, it gives intense crumb formation and weed control over the whole surface. It may still be used when the plants are hand-high, but there is a risk of virus transmission to injured plants. The problem is largely avoided by fitting a ridge weed harrow unit to the ridging body (ridging weed harrow).

When plant growth is in the more advanced stage, vehicular traffic should be kept to a minimum. Lateral pressure from tractor wheels compacts the soil in the ridges and this inhibits growth in the tubers. Haulms are more resistant to fungus and insect attack if c. 30 kg/ha (28 lb/acre) of calcified seaweed is dusted on to the leaves. Fungus attacks may be treated with 3–5% sodium silicate (waterglass) or with Bio-S.*

---

* Bio-S was developed in the 1950s for fruit growing in the harsh climate of the Swabian Alb in Germany. It contains a number of herbal extracts combined with sulfur. Bio-S is used as a universal remedy against fungal attack, generally in 0.5–1% solution.

If Colorado beetles are found, ladybirds may be utilized for biological control. A high ladybird population is achieved by providing habitats for them – woodland, hedges and baulks. Then aphids have to be present to attract the ladybirds to the potato field where they will mate and lay their eggs. The larvae will feed on the eggs of the Colorado beetles that arrive at a later date, greatly reducing the extent of the damage.

In an emergency, with a large Colorado beetle population and too few ladybirds present, Spruzit (liquid or dust) may be used. It is only really effective against the young larvae. (In the UK and Ireland, Colorado beetle is a notifiable pest.)

Dynamic measures

Potatoes react quite noticeably to dynamic measures. Thus one-sided trends develop if all soil preparation, planting, cultivation and spraying (to stimulate root and tuber development) are done at a root trine (or trigon), as determined by the relative positions of moon and zodiac (see Figure 22). Experiments of this kind done at Talhof Farm have shown the following:

1. Stem and leaf development tended to be weak and stunted, so that in spite of a well-developed root system and a good number of tubers per plant, poor assimilation meant that the yield was not satisfactory.
2. Perennial weeds such as bindweed, couch, thistles and lucerne initially produced only limited aerial growth but extensive root systems. Late summer then brought vigorous growth, little impeded by the limited shade produced by the potato haulms. The result was late weed development in the field.

> Using a horse for cultivation work avoids compacting the soil and consequent losses in yield.

9. ARABLE CROPS

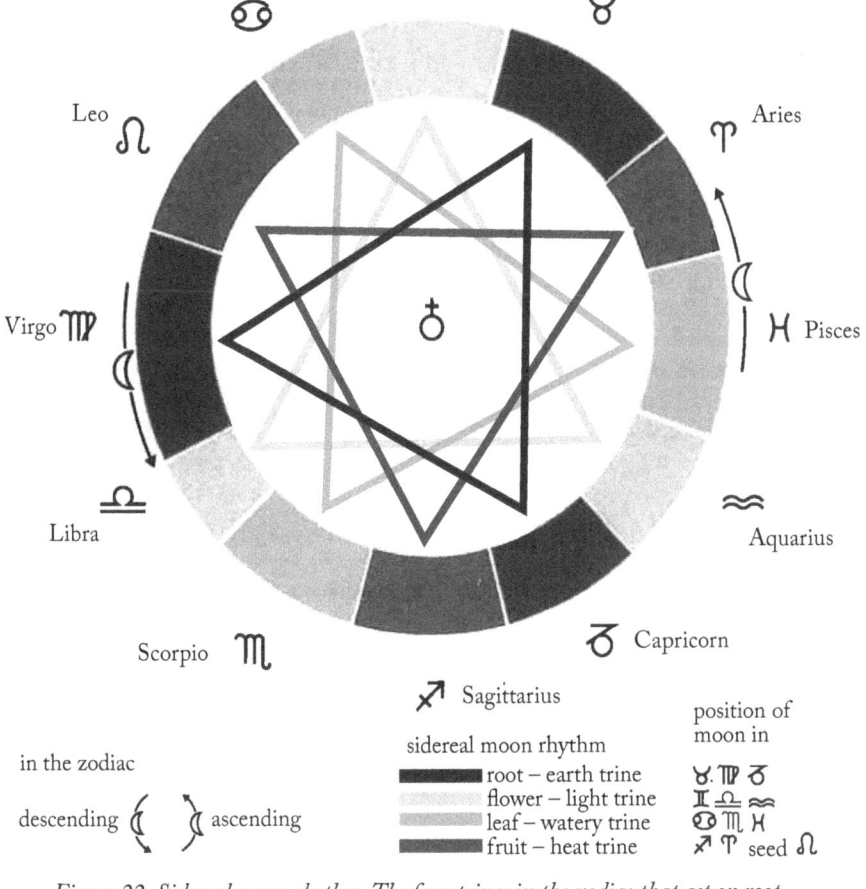

Figure 22. Sidereal moon rhythm. The four trines in the zodiac that act on root, leaf, flower and fruit. Based on the work of Maria Thun.

We therefore do not limit operations to root days,
- but also work at leaf trines during the main period of foliage growth (horn silica sprayed in the morning, hoeing and ridging) and
- at seed and fruit trines to support the ripening process (horn silica in the afternoon).

Figure 23 shows a work scheme to illustrate the method.

The scheme merely serves as a general guide. As always, soil condition, plant development and weather conditions are important aspects that must be given prime consideration with every operation.

On sites where there is little risk of degeneration and fungus attack, birch pit concentrate may be used for the seed bath. It will usually give increased yields but also increases susceptibility. The valerian preparation is better on heavy soils, the oak bark preparation if there is a high risk of degeneration and fungus attack. Seed baths should as far as possible be applied at root trines. In humid climates that encourage

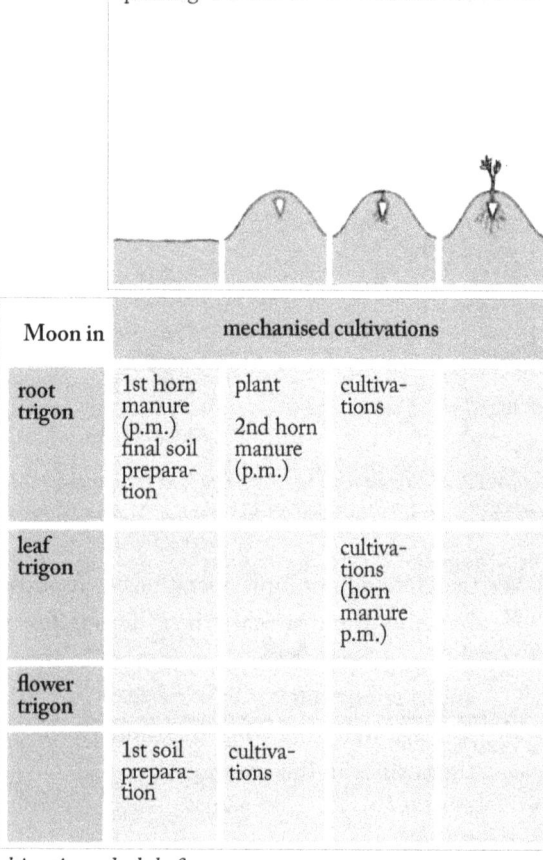

Moon in	before planting	before emergence	emergence
root trigon	1st horn manure (p.m.) final soil preparation	plant 2nd horn manure (p.m.)	cultivations
leaf trigon			cultivations (horn manure p.m.)
flower trigon			
	1st soil preparation	cultivations	

Figure 23. Sample cultivation schedule for potatoes.

9. ARABLE CROPS

prolific growth a single application of horn manure will suffice. Potatoes respond well to frequent applications of horn silica.

In sudden and extreme weather changes, e.g. from hot and dry to damp and warm, horn silica applied on three consecutive days gives good results. This certainly was the Talhof experience in 1983, a dry year. Our potato crops were completely fungus-free. Colorado beetles did not attack, though on other farms in the area insecticides had to be used repeatedly to get them under control. The biodynamic field sprays should not be regarded as direct control measures; they have an indirect effect by strengthening crops and making them unattractive to pests and diseases.

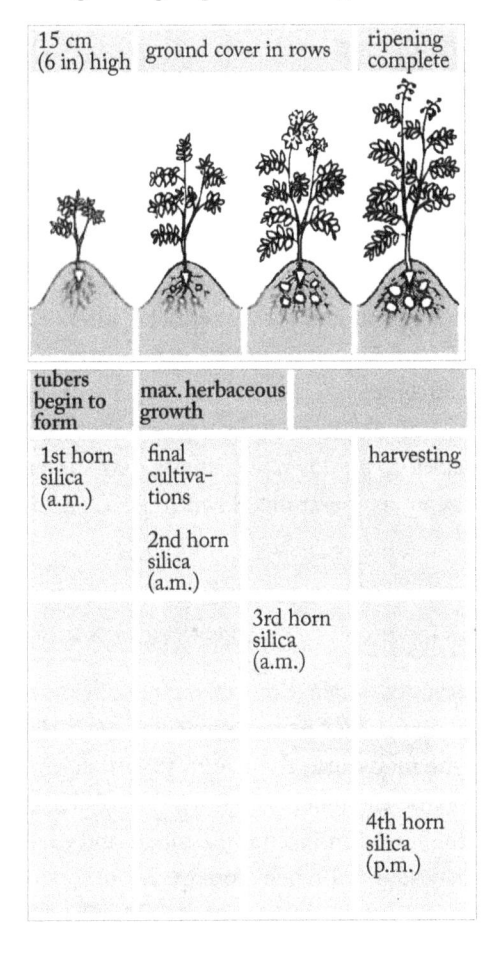

15 cm (6 in) high	ground cover in rows		ripening complete
tubers begin to form	max. herbaceous growth		
1st horn silica (a.m.)	final cultivations		harvesting
	2nd horn silica (a.m.)		
		3rd horn silica (a.m.)	
		4th horn silica (p.m.)	

risk periods		rain mm	changes in weather conditions: max./min. temp. °C	rel. humidity %
1)	July 7	5.9	12.5/27.0	86
	July 8	7.1	16.0/27.5	92
	July 9	0.0	13.5/30.0	89
2)	July 20	1.8	15.0/26.5	81
	July 21	0.0	6.5/23.5	68
	July 22	0.0	4.0/29.5	68
3)	Aug 9	0.0	8.0/27.8	86
	Aug 10	8.8	8.0/28.0	91
	Aug 11	0.2	10.0/25.0	93

Neither leaf nor tuber blight developed. Colorado beetle did not attack.

Table 25. Example of horn silica applications on potatoes to increase resistance in relation to changes in weather conditions. In damp warm weather or risk of this, potatoes are sprayed on three consecutive days (always at 7 a.m. true local time).

Horsetail tea is sprayed in autumn and spring on fields intended for potatoes to prevent fungus attacks. Spraying on three consecutive days has proved particularly effective.

If there is a risk of late or early frost, we spray valerian the night before temperatures are expected to go down. Experience has shown that this will prevent damage at temperatures down to −3°C, or at most −4°C (27° or 25°F).

Harvest

Potatoes should only be harvested when the tubers are fully ripe. With most varieties, the moment of ripeness coincides with the withering of the leaves. Blight may cause the haulm to wither and die in just a few days. In that case the tubers need to be harvested as soon as possible; otherwise the infection reaches the tubers and even healthy looking specimens may soon rot when stored.

As already mentioned, plants grown from specially chitted seed stay green longer. In this case the tubers are tested for ripeness. They separate easily from the shoot, the skin is firm, and when a medium-sized potato is cut in half the cut surface is dry.

Potatoes harvested at a root trine normally keep better, store better and show better dormancy. This has been shown in experiments carried out by Maria Thun. However, if the other criteria – physiological maturity, dry soil and weather conditions – are met, it is not advisable to wait for a favourable moon position at this late season.

After the harvest, potatoes are left to rest and dry for one or two weeks before they are sorted and put into winter storage.

Sugar beet

Sugar beet *(Beta vulgaris* ssp. *vulgaris* var. *altissima)* is not normally grown as a biodynamic crop as it makes tremendous demands as regards soil, climate and manuring. Separate sugar production would also mean making very large deliveries to sugar factories. Because of the physiology of nutrition, there also is no demand for refined sugar from biodynamic sources. The sugar beet required for the manufacture of Demeter Syrup is grown by the method used for fodder beet.

Fodder beet, mangels and swedes

Fodder beet and mangels *(Beta vulgaris* var. *alba)* and swedes *(Brassica napus* var. *napobrassica)* are considered an excellent, palatable succulent feed as part of the winter rations. Being eminently digestible and high energy-yielding, they may be used as a protein balancer in dairying. It has been established that milk protein levels may be enhanced by feeding fodder beet.

Labour requirements are relatively high if no herbicides are used, adding to the peak requirement created by the first cut for fodder. If the fodder beet area is relatively large, therefore, the necessary cultivations may compete with silage cutting and haying. It is wiser not to grow beets if beet cultivation will not leave enough time for the fodder harvest; top quality roughage has absolute priority with cattle. However, even small amounts of beet may be of benefit in so far as they improve the digestibility of the whole ration. Fodder sugar beet makes an excellent addition to the ration for horses and sows.

A wide range of varieties is available so that fodder beet can be grown in almost any soil and climate. Varieties with a larger proportion of root in the soil are more suitable for light, relatively dry soils as they will do better during dry periods. Swedes do particularly well in cooler climates with high humidity, as do fodder beet or mangels, which have a higher proportion of root above the soil; these also do better in heavy soils.

Suitable *preceding crops* are above all long leys and potatoes, as the need for weed control operations is reduced.

Leys need to be ploughed up in good time, producing a deep summer furrow; root residues will then not interfere with hoeing operations and it is possible to have two or three weed clearing operations. If it should be necessary to plough again in autumn, the furrow needs to be shallow to prevent bringing up new weed seed. Phacelia (California bluebell) may be useful as ground cover and green manure. It is helpful to manure with humified composted manure or well fermented liquid manure before shallow ploughing-in when the plants are about 10–20 cm (4–8 in) high.

Fodder beet responds particularly well to one or two applications of dung liquor. The best method is to use a liquid manure drill and make application between rows on slightly damp soil when the plants are at the right stage of development.

When preparing the seed bed in spring it is important
- not to loosen settled top soil, but work to it shallowly,
- to take special care to protect the soil water, whatever the operation,
- to allow as many weeds as possible to germinate and remove them before drilling.

Floats, culti-harrows and seed bed combinations may be used for this purpose.

Spaced drilling (using pelleted or monogerm seed) with the drills designed for that purpose greatly reduces the labour required for subsequent cultivations. Row spacings of 50 cm (20 in) and within-row spacings of 15–20 cm (6–8 in) have proved effective in producing the optimum yield of 80,000–100,000 roots per hectare (30,000–40,000 per acre). Within-row spacing should be reduced where there is a risk of pygmy beetle or wireworm attack.

Depending on soil and weather conditions, the first pre-emergence cultivation is done blind, using a weed harrow, harrow, mechanical hoe or flame weeder. The wheel marks left by the drill provide the necessary orientation. It may sometimes be necessary to use the roll to reestablish soil closure during a dry spell or break up puddled or crusted top crumb.

Beet and swede cultivation without the use of herbicides competes with silage and hay making; only limited field space is therefore given to these crops.

Subsequent cultivations serve to provide good growing conditions for the crop until ground cover in the rows is complete, and tilth will then be created and maintained in the leaf shade provided by the crop itself. If crops can be kept weed-free until then, late weed growth will normally be prevented right up to harvest time. It is important to see that the upper crumb layer between rows is always kept loose by timely hoeing using angled and duck's foot shares, initially with leaf protectors fitted. This makes it difficult for weed seeds to germinate, whilst the roots of the beet crop are given good growing conditions. In sufficiently humid weather conditions where the soil has plenty of life in it, the soil is covered with a dense felt of fine white rootlets. High yields are achieved in this way and the soil is left in good condition for following cereal crops.

Swedes, and fodder beet, too, are good crops to plant providing there is the prospect of rain when planting is done, or the field can be watered with sprinklers. The method permits full utilization of the good preceding crop effect of a dense winter catch crop of fodder plants (e.g. Landsberg mixture), which leaves the ground weed-free. Soil structure and nitrogen enrichment provide excellent conditions for the seedlings.

The camomile preparation is used as a seed bath for swedes, and birch pit concentrate made up with diluted whole milk for fodder beet. As with potatoes, rhythmical applications of horn silica made early in the morning on three consecutive days strengthen the crop to withstand extreme weather conditions. Pests are usually kept under control by spraying with 24–36-hour-old nettle liquor.

Pelleted seed without added fungicides and insecticides is now used widely. Pelleted seeds of severeal species is produced by the Bingenheimer Saatgut.

Beetroot

Like the carrot, beetroot *(Beta vulgaris* ssp. *vulgaris* var. *conditiva)* is important as a food and for health. Raw salads made with beetroot, apples, celeriac and parsnip or beetroot juice increase human resistance to disease. Where beetroot juice is much in demand, this is an important field vegetable.

Beetroot and fodder beet both belong to the goosefoot family (Chenopodiaceae). The requirements for soil preparation are about the same. In a cropping sequence beetroot will still do well if grown as a second crop; potatoes are the best preceding crop. As with other beets, the 'fruit', i.e. the part that is harvested, develops through thickening of the hypocotyl (the part between stem and root), and may store excessive amounts of nitrate (NO_3). Only well rotted composted farmyard manure should be applied directly to ensure good flavour and storage qualities; the method of choice will always be to manure the preceding crop.

For juice production on an industrial scale, beetroot is grown as a main crop, with 4–6 kg/ha (3½–5½ lb/acre) drilled in rows spaced at 50 cm (20 in), and within-row spacing at 6–10 cm (2½–4 in). Drilling is done relatively late, when the soil has warmed up, so that germination and growth will proceed apace. This reduces the period when the young plants are susceptible to damping off (due to a fungus) and pests such as pygmy beetle.

Large roots are required for industrial processing, smaller ones for sale as a vegetable. Smaller beetroot is grown as a second crop after winter catch crops, leeks or lettuce, the within-row spacing being 4–6 cm (1½– 2½ in). In favourable sites beetroot may still be planted after early potatoes, winter rape or vegetable peas. It is important not to remove the root tips as this may delay growth and result in fanginess.

The leaves and the cake left after juicing make a valuable supplementary cattle feed.

Carrots

The carrot *(Daucus carota* ssp. *sativus)* belongs to the family known as the Umbelliferae. Its tradition as a major vegetable crop in Europe and Asia goes back over more than 4,000 years, and it has tremendous importance as a food for young children as well as fodder for young animals.

The intensely coloured root is rich in carotene, indicating that in this plant the 'flowering' and 'fruiting' processes have moved down into the root region, whereas in most plants they take place under the influence of light and heat in the flowering region. All plants produce carotene (vitamin A) under the influence of light in their green parts, but the vitamin is soon broken down again when the plant dries in the sun. In the human and animal organism, carotene is converted to vitamin A if there is sufficient exposure to light. The vitamin is important for good eyesight and for the function of the skin and all mucous membranes. Deficiency may result in night blindness; in animals it may cause fertility problems and increased susceptibility to infections of the respiratory and gastrointestinal tracts such as influenza and diarrhoea.

Carrots grow best in deep, medium and humic soils with an adequate supply of lime.

As to climatic conditions, extended dry periods in spring can be a problem; prolonged droughts in July and August, when the tap root is developing to its full extent, may cause deformation. Subsequent rain or the use of sprinklers will sometimes cause the carrots to split. The crop is relatively frost resistant, taking no harm in quite low temperatures in autumn. Mild and warm weather periods in late summer and autumn allow the roots to ripen fully, giving high, top quality yields.

Carrots are best preceded by crops that leave the soil loose, with a good crumb and largely weed-free. Potatoes are good, as are cereals that can be cleared early, so that further weed control operations are possible after harvest. On heavy clay, carrots can be grown after a long ley, providing this is ploughed in sufficiently early.

Direct manuring has an adverse effect on the quality of the crop, which assimilates nutrients easily. The only exception is fully humified composted horse manure; up to 20 t/ha (9 tons/acre) will give high yields, sound plants and top quality. Apart from this, manuring should be limited to the preceding crop.

Cultivations aim to prepare a firm seed bed, with garden-fine tilth at the surface, free from subsoil and underground consolidation. Although seedlings are well able to withstand late frosts, it is advisable to drill late – in the second half of April but no later than mid-May, as the roots may not ripen completely with a later drilling date. Final preparation after use of a seed bed combination or the like consists in a pass with the roll; this will make hoeing easier at later stages. Field crops for food are drilled in rows spaced at 35–50 cm (14–20 in), using about 2 kg seed/ha (2 lb/acre); with carrots for industrial processing, rows spaced at 50 cm (20 in) and a maximum of 1.5 kg/ha (21 oz/acre) have been found ideal. Damage due to pressure from tractor wheels is best avoided by spacing rows further apart where the wheel tracks are. Fanginess is increased in areas where spring cultivations have consolidated the soil.

Precision seed drilling with pressure wheels will permit blind cultivations and make crop cultivation easier.

The seed frequently takes three or four weeks to germinate. Seed baths with birch pit concentrate shorten this period by about a week. It then takes several weeks until inter-row ground cover is complete, with the seedlings highly sensitive to competition from weeds. Pre-emergence weed control is essential, flame weeding being highly effective. Implements may be suitably combined to effect blind harrowing and flame weeding in one pass immediately before

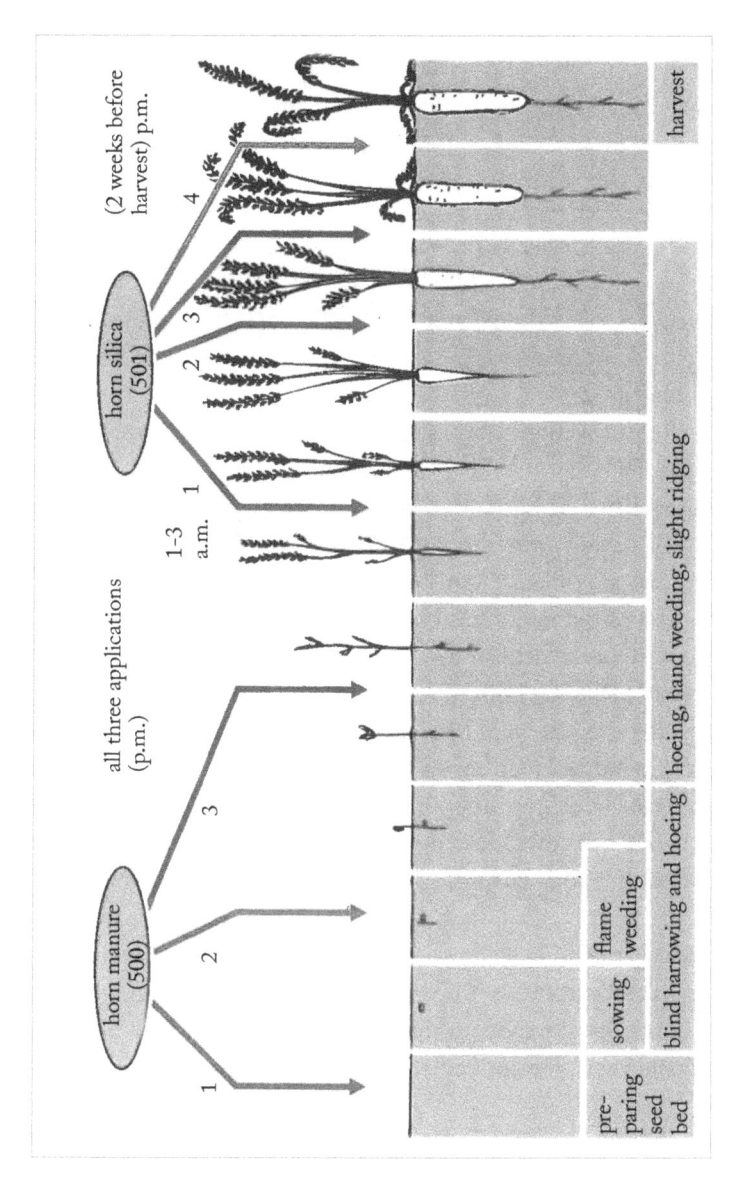

Figure 24. Cultivations required at different stages of plant growth for carrots.

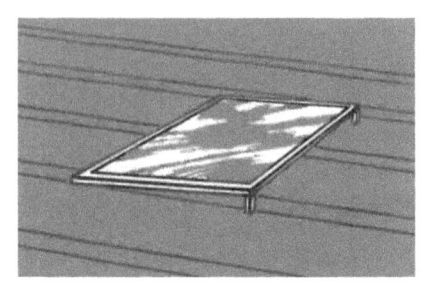

Figure 25. A piece of glass put over the rows after sowing accelerates germination thus helping to find the right time for pre-emergence flame weeding.

emergence. Crusts should not develop, as they tend to break up on hoeing and move sideways, damaging seedlings even if protectors have been fitted.

Further cultivations start as soon as the seedlings appear. Machine hoeing comes first, working as close to the rows as possible, with protectors fitted. It is followed by hand weeding. Quick and efficient work will usually win the battle against wild plants. As the crop grows, rows are ridged slightly by putting the duck's foot shares at an angle. Less of the root then shows above ground and carrot flies have less opportunity to deposit their eggs. Singling has its dangers for this very reason. Ridging also prevents the tops of roots from greening.

The first application of horn silica is made when the foliage is about hand high and the root begins to thicken. Before that, horn manure is applied two or three times. Two or three more applications of horn silica before the rows close up will enhance the sweet and delicate flavour of the roots. Further applications made in the mornings may cause hardening and bitterness. In early autumn, when the roots develop their main bulk, horn silica is sprayed in the afternoons about six weeks and then again three weeks before the proposed harvesting date. This helps the roots to ripen fully in a balanced way, improving nutritional value, flavour and keeping quality.

White and red cabbage

Cabbages, like swede and rape, belong to the large, widespread and vigorous crucifer family (Cruciferae). White and red cabbage *(Brassica oleracea* var. *capitata* f. *alba* and f. *rubra)* are the most likely to be grown as field crops. Sauerkraut (lactic fermented cabbage) is particularly important in human nutrition. It is digestible and promotes health and vitality.

Cabbages are generally considered to be rather indigestible, putting a strain on the metabolism. This is not the case, however, with cabbages grown as part of a rich and varied cropping sequence, using balanced manuring techniques that will not force growth and regular applications of biodynamic field sprays. Again and again we have found that customers who had found cabbage quite indigestible before keep coming back for Demeter quality cabbages and sauerkraut, having found them perfectly digestible and highly palatable.

White cabbage requires about the same kind of soil and climatic conditions as swedes, except that more moisture is needed. Red cabbage is more exacting with regard to soil quality and prefers a milder climate.

All crops are suitable as preceding crops except for members of the crucifer family, which are subject to the same pests and have similar requirements.

Vegetable seed of biodynamic origin is frequently superior to conventionally produced seed. Crops will be healthier and more resistant, and this applies particularly to cabbage seed.

Seedlings for main crops are best grown in the field intended for the crop itself. Spacing the rows 50 cm (20 in) apart permits machine hoeing and produces strong, resistant plants. A well developed main stem acts as a water reservoir. Planting is done in a well prepared, weed-free field; once the plants have come upright, a mechanical hoe may be used, creating slight ridges. One or two more passes in which the ridging is increased will make hand hoeing largely unnecessary.

Cabbages grown for storage are best given individual spacing of 33 by 75 cm (13 × 30 in). This reduces leaf damage and allows cultivations to be done also at a relatively late stage. Yields will be no less than with narrower spacings.

No action needs to be taken with aphids if there is a good *Trichogramma* wasp population and the crop is in good condition. Care should be taken, however, to overcome the inhibition of growth that has given rise to the attack, e.g. by using sprinklers to overcome dryness. Occasional stunted plants will retain sufficient aphids to provide food for the parasites and prevent a further massive attack.

At Dottenfeld Farm (Dottenfelder Hof) 1976 was a dry year when approximately every plant in a 1.5 ha field had two sets of cabbage white eggs. The resulting caterpillar population was horrendous, eating away the leaves from the margins. When the caterpillars were about 2 cm (1 in) long, every single one died – they had been attacked by the wasp. The harvest was good, yielding top quality cabbages.

Massive attacks from cabbage fly, cabbage flea and cabbage stem weevil on young plants need treatment with Spruzit (a contact poison containing pyrethrum that is safe for bees, animals and humans); older plants are not threatened by them. Pests often prefer nearby crucifers such as charlock and pennycress and will only move on to the vegetable crop following their removal.

Timely applications of freshly made nettle tonic will increase crop resistance and inhibit the development of the pests.

Clubroot, a fungus disease *(Plasmodiophora brassicae)*, may cause serious losses. The only way of combating this is to keep crucifers well apart in cropping sequences and make sure the soil has sufficient lime (pH not less than 7). The fungus attacks all crucifers and may persist for years on the arable weeds belonging to the family.

9. ARABLE CROPS

Legumes

All pulses belong to the pea family (Leguminosae), whose 12,000 species grow in all parts of the globe. In Europe 20–30 species are grown as agricultural crops.

Pulses have always been important providers of protein for both humans and animals. Vetches and lentils are known to have been in existence as early as 4000 BC. In the Old Testament we have the story of Isaac's son Esau selling his rights as the first born to Jacob for a dish of lentils. Diocles (a fourth century BC Greek physician) and Hippocrates wrote that the hyacinth bean or lablab black bean *(Dolichos lablab)* was widely eaten in Italy. This bean still provides the basic protein for people and animals today in many parts of India, South America and Africa. It contains sufficient lysine (an essential amino acid) to prevent protein deficiency diseases. Other types of beans and lentils and peas also serve as foods.

Legumes have the important ability to fix atmospheric nitrogen through nodule-producing bacteria that live in symbiosis with them *(Rhizobium)*. They are known as 'nitrogen accumulators'. They are also well able to dissolve soil minerals, thus leaving a topsoil rich in available nutrients for the following crop. The pea family are therefore classified as nitrogen assimilators and enrichers as well as nutrient processors, compared to most other plants which consume and reduce both nitrogen and mineral stores.

> Legumes are able to enrich the soil with nitrogen from the atmosphere fixed by nodule-producing bacteria on their roots.

The most important grain legumes are the following large-seeded species:
- peas *(Pisum)*
- vetches *(Vicia)*
- vetchling *(Lathyrus)*
- lupin *(Lupinus)*

143

- beans *(Phaseolus)*
- lentils *(Lens)*
- soya beans *(Glycine)*

They are grown for grain to produce pulses for human consumption and animal fodder and as green fodder or green manure, sometimes as main, but generally as catch crops.

Small-seeded legumes are major field fodder plants, together with grasses and members of the crucifer family, with some of them also extremely valuable in leys:

- lucerne/alfalfa *(Medicago)*
- clovers *(Trifolium)*
- birdsfoot trefoil *(Lotus)*
- serradella *(Ornithopus)*
- sainfoin *(Onobrychis)*
- melilot *(Melilotus)*

Peas

Both garden peas *(Pisum sativum)* and field peas *(Pisum arvense)* are grown, the first mainly as a vegetable but also for large-grain fodder grain, the second mainly as green fodder and green manure and to produce fodder grain. Many different varieties are available so that the right kind can be found for almost any site. Peas grow best in mild sites rich in humus and in warm places with medium soil. Clays and loamy sand with good lime and humus levels are still just possible for peas; heavy clay, lean sandy and low-lime boggy soil are not suitable. Field peas include some less demanding varieties, such as the grey pea, which will also give reasonable results in less desirable sites, providing there is sufficient lime.

Field peas are also less demanding concerning climate. Peas generally like dry, warm sites best and need adequate moisture when flowering in May or June. Very high temperatures and drought prevent the seed from setting; more than half the flowers may wither and drop off. The plants cope well with occasional late frosts; prolonged cold spells in spring with frequent night frosts will, however, seriously delay and inhibit growth. It is better to drill later in such areas, though generally speaking drilling should be done quite early.

A camomile seed bath is highly effective. It accelerates germination and the development of root nodules. Repeated field trials have shown that plants from untreated seed are slower to develop and have fewer nodules.

The more demanding varieties produce less foliage and therefore need better weed control. They give better yields when following on row crops, potatoes or root crops than on cereals. Crops preceded by weed-free cereal crops and a crucifer family summer catch crop may however give equally good yields.

Crucifers and peas help one another. Some added mustard *(Sinapis alba*, 0.7 kg/100 kg peas) provides excellent support for peas. The mustard seed is only mixed in each time the hopper is refilled during drilling, otherwise it will separate. Just sprinkle it over the peas; it will mix in on its own.

A good seed bed is important, with the first weed generation removed before drilling the peas. Depending on vigour and local conditions, rows are spaced 20–30 cm (8–12 in) apart. Depth of drilling depends on the size of the seed – about 3 cm (1 in) for small seed and up to 8 cm (3 in) for larger ones. It is possible to put the peas in deeper if there is a risk of pigeons or crows eating them, and this also makes subsequent cultivations easier. Depending on growing conditions, 700–900 times the thousand grain weight (TGW) is drilled. Too little seed will mean gaps in the field. For instance, a TGW of 300 g will require 240 kg/ha:

$$\frac{300 \text{ g} \times 800}{1{,}000} = 240 \text{ kg/ha}$$

Peas are relatively slow to develop and need special attention (competition from weeds). It is generally helpful to use a light harrow until the peas are 10 cm (4 in) high and a mechanical hoe after this. Once the peas begin to develop tendrils further cultivation operations would cause damage.

On difficult sites it has proved useful to add peas to cereals rather than grow them on their own. Add about 20 or 30 kg/ha (18–28 lb/acre) of peas to the normal amount of oat, wheat or rye seed. The main crop will give the usual yield, and an appreciable number of peas are an extra gain. In unfavourable sites the method will at least produce sufficient pea seed for green manuring and fodder. A helical grader will separate the peas from wheat and rye.

> Peas and field beans grown for grain provide good farm-grown protein to add to the winter rations for dairy cows.

The valerian preparation will markedly improve flowering capacity and therefore yields. It is sprayed the first time when the plants are 10–20 cm (4–8 in) high and a second time just before flowering.

The varieties now available ripen evenly and are therefore all suitable for combine harvesting. Sufficient peas are generally dropped to produce a good green fodder or at least green manure crop to follow. Even a heavy dew will cause the seeds to swell up; it is therefore advisable to bring in the straw quickly and work the stubble with a disk harrow or other suitable implement.

In a favourable climate peas may be followed by winter rape, winter barley or winter wheat. A successful pea crop will leave them a clean, fertile soil with good crumb structure.

Field beans

Field beans *(Vicia faba)* need more water but are not as dependent on warmth as peas. Sites with high atmospheric humidity and deep, well structured soils are ideal. This crop has poor tolerance for heat during the flowering period. Black bean aphid is liable to attack as well, reducing yields even further. Two applications of valerian will help with flowering.

Well-rotted composted farmyard manure applied in early autumn after the preceding crop (generally cereals) will enhance the development and preceding crop value of the beans. Early maturity and timely clearance – not always possible – provide good conditions for winter wheat as a following crop.

Drilling is done as early as possible at a depth of 6–10 cm (2½–4 in), depending on size. Rows spaced at 25–35 cm (10–14 in) permit mechanical hoeing, with the crop responding well to this. Post drilling cultivation until the plants are 15–20 cm (6–8 in) high is done with harrow and weed harrow, moving at a speed to suit the stage of development and soil conditions. As already mentioned, there is a risk of late weed development, as the lower leaves die off during the relatively long time the seeds need to mature. About 10% of fodder peas added to the seed provide the necessary ground cover and prevent weed growth.

> Small-seeded field bean varieties may also be used in green fodder mixtures.

Vetches

Common vetch *(Vicia sativa)* is mainly grown for green fodder. The grain can be used for feed but the grain yield is too low to make this really worth while. The crop is less demanding than field bean and pea but sensitive to dry periods, which may result in a poor crop. On the other hand it does do well even on tough clayey soils.

Growing for grain is best done together with late maturing oats, beans and oats, or spring wheat. A useful mixture has proved to be 120 kg/ha (110 lb/acre) of vetches and 60 kg/ha (55 lb/acre) each of beans and oats. Varieties should be chosen that mature all more or less at the same time. Seed required for growing fodder can be added (10–20 kg/ha, 9–18 lb/acre) to oats of spring wheat. Vetches tend to ripen unevenly and it is therefore helpful to windrow as soon as the lower pods begin to turn brown. Leave to dry for a day and turn carefully before picking up with the combine the next day. The grain usually needs additional drying (risk of mould attack).

Hairy vetch *(Vicia villosa)* and Hungarian vetch *(Vicia pannonica)* are grown mainly as winter catch crops.

Lentils

The lentil *(Lens culinaris)* is generally considered the finest of the pulses. Large-seeded varieties grow best in light to medium soils or in shallow soil on limestone parent material. Lentils do not tolerate the wet well and do better in warm, dry weather, especially from flowering to maturity. They are highly sensitive to competition from weeds, which is why they are often grown after potatoes – sometimes mixed with early maturing spring barley. 100 kg seed/ha (90 lb/acre) is combined with 60 kg/ha (55 lb/acre) of spring barley; with lentils only one needs 150 kg/ha (135 lb/acre). Spacing between rows is 18–20 cm (7–8 in), depth 3–5 cm (1–2 in). Weed control is necessary before and immediately after drilling. Lentils do not tolerate harrowing or weed harrowing once they have emerged. Before the young plants begin to entwine, a pass is made with the mechanical hoe. Lentils can be windrowed like vetches; they are turned and only picked up with the combine when fully ripe.

The camomile preparation has proved most effective as a seed bath, as with all legumes. Lentils and vetches are treated with valerian by the method given above for peas.

Healthy lentil straw harvested when dry is approximately equal to medium quality hay in its value as animal fodder.

Lupins

Distinction is made between bitter and sweet (low in bitter compounds) lupins. Bitter lupins can only be used for green manuring. They do not make suitable fodder because all parts contain the bitter compound lupinidine. The most important species are yellow lupin *(Lupinus luteus)*, blue lupin *(L. angustifolius)* and white lupin *(L. albus)*. Sweet lupins provide highly palatable green fodder and protein-rich feed grain. All varieties have long tap roots and are therefore used as a pioneer crop for light soils (white lupins also perform this function in heavy soils). Soil fertility is given a long-lasting boost, and lupins make the ideal preceding crop for cereals, row crops and above all potatoes. They do require a fairly warm climate, otherwise development tends to be delayed in the early stages. Lupins tolerate late frosts with temperatures down to -3 or -4 °C (27 or 25°F).

During germination and at the juvenile stage lupins need adequate moisture; they are drought-proof once the tap root has fully developed. Dry weather is desirable for the flowering period. Blue lupins need some atmospheric humidity even then, otherwise some of the flowers will drop. Sites more than 300 metres (1,000 ft) above sea level are not suitable for seed production.

Yellow lupins prefer light sandy soils and are highly sensitive to lime. The pH should not be less than 4.3, however.

Blue lupins are slightly more exacting, with seed production requiring soil of the same quality as that needed for oats; sensitivity to lime, on the other hand, is slightly less.

White lupins definitely need better quality soil such as clay or loess. They are more tolerant of lime than other lupins but also will not grow in weathered limestone soil. In the Central European climate white lupin seeds ripen very late, if at all, compared to end of July to early August for blue and middle to end of August for yellow lupins.

Adequately wilted, lupins mixed with oats and peas provide valuable silage material.

With initial development slow, the field must be free of perennial weeds. Annual weeds can be controlled with harrow and weed harrow after drilling.

The amount of seed required is 150–180 kg/ha (135–160 lb/acre) for yellow varieties, 160–200 kg/ha (145–180 lb/acre) for blue, and 200–250 kg/ha (180–225 lb/acre) for white lupins. The depth should be not more than 4 or 5 cm (1½–2 in) as the cotyledons are brought above ground and therefore do not tolerate being covered too deeply. Germinating power tends to be on the poor side (hard seed coat), so that a germination test is essential. Rows spaced at 20–30 cm (8–12 in) have proved effective.

Mechanical hoeing starts as soon as rows of seedlings are visible, using protectors and repeating the operation as required until the lupins cover the soil. Later on the dense crop will suppress weeds completely.

Lupins are self-pollinating, though yellow varieties also tend towards cross pollination. Care must thus be taken when growing for seed that no other variety is growing nearby, especially no bitter lupins (not even for green manuring). The first time lupins are grown the seed needs to be inoculated with bacterial cultures *(Bacterium rhizobium lupini)*. Damage by game tends to be high in fields close to woods where the game population is high. It may prove necessary to put up fencing. The pods of modern varieties are burst-resistant so that a combine may be used for harvesting. Further drying will be necessary.

Miscellaneous crops

Winter rape

In conventional farming, winter rape *(Brassica napus* var. *napus)* is considered a good cleaning crop in cropping sequences with a high proportion of cereals. It leaves an excellent crumb structure and is one of the best preceding crops for wheat.

Rape oil of biodynamic origin has only a very small market, which is why rape is limited to a few particularly suitable sites and to large farms. The information given below comes from a large farm in the Cologne/Aachen area of Germany.

Rape is very demanding with regard to cultivation, climate and manuring. It needs to achieve considerable growth very early in the year, before biological activity starts in the soil. Rape is therefore considered a difficult crop. Suitable preceding crops are legumes, one or two year leys, or peas grown for grain. Rape grown after cereals requires plenty of composted manure to which commercially available organic fertilizers such as pigs' bristles has been added. This is not advisable, however, as the use of such compost is apt to result in massive pest attacks, with serious damage caused by rape flea, aphids and blossom rape beetles.

The seed bed needs to be prepared at least four weeks before drilling to allow the ley turves to rot down sufficiently. An early second cut and an application of composted manure (15–20 t/ha, 6½–9 tons/acre) are followed by shallow rotavation and later by grubbing. Repeated applications of Maria Thun's cowpat preparation enhance the conversion activity in the soil.

On lighter soils drilling may be done immediately after ploughing, with packer, and fine seed bed preparation, the seed having been dusted with wood ash to prevent rape flea damage.

Development tends to be rapid, so that three applications of horn silica can be made in autumn, always in the early dew.

If there are many weeds whilst the rape is in the early growth stages, these are easily removed by mechanical hoeing in rows spaced 30 or 40 cm (12–16 in) apart. Later the crop will suppress all weeds including thistles, since it is cut while still at the flowering stage, using a windrow combine. A seed furrow for the following stubble-sown catch crop enhances the effect.

Blossom rape beetle (*Meligethes aeneus*) may attack in spring. The local climate is mild, with winter rape coming into flower early, so that there is little damage. Flowering will be more rapid if there are sufficient beehives placed on the baulk; the beetles generally disappear soon, and increased yields may be expected. After a one-year ley as preceding crop, yields of erucic acid-free varieties are in the region of

2.4 t/ha, after a two-year ley about 2.8 t/ha (1.1 and 1.25 ton/acre).

Summer rape generally yields about 0.5 t/ha (0.2 tons/acre) less. Flowering is later so that blossom rape beetle is often a problem. Summer rape is not to be recommended therefore, or at least only for fodder.

Flax

Linseed is important in the rearing of young stock, as a feed for horses and a medicine. Flax *(Linum usitatissimum)* is therefore a useful crop for home consumption.

Linseed is grown from special varieties that require less moisture than flax grown for fibre. They like a fairly warm climate and grow in almost any kind of soil except on poorly drained or extremely lean sites. Deep loams are best in dry regions.

Varieties grown for fibre need a good water supply and a weed-free field; otherwise cultivations have to be frequent. Flax harvesters do not function well if the crop includes many weeds.

Both types of flax will grow after any preceding crop, but row crops are best for flax grown for fibre because they leave the field clean. For good quality fibre the flax needs to be densely grown, and after legumes there is a definite risk of the crop becoming lodged. Legumes are therefore better before flax grown for linseed.

The seed bed must be of garden quality and firm. Drilling is done at a depth of 2 or 3 cm (1 in) in the first half of April. An old saying is that flax needs to be drilled on the 100th day of the year (10 April) and can be pulled on the 200th (20 July). Rows are spaced at 12 cm (5 in), up to a maximum of 18 cm (7 in); 18–25 cm (7–10 in) for linseed. Flax grown for fibre needs 120–150 kg seed/ha (110–135 lb/acre), or ideally 2,000 plants to the square metre, linseed 70–100 kg/ha (60–90 lb/acre, 1,000 plants to the square metre).

The only suitable manure is humified composted manure, preferably made with horse manure. It is spread at the end of the winter and worked in shallowly when preparing the seed bed. This will accelerate development in the early stages and avoid the risk of the crop becoming lodged or being seriously attacked by flea beetles.*

Cultivation is limited to one or two passes with the mechanical hoe for linseed. Fibre varieties need hand hoeing in addition, and later on tall weeds liable to be a nuisance have to be pulled by hand. Hoeing operations must be finished by the time the plants reach a height of 20 cm (8 in).

Linseed is combine harvested when dead ripe. The straw may be used to produce coarse linen goods like sacking or oakum. Fibre varieties have to be pulled, either by hand or, in larger fields, by machine. The stems should be a yellowy green at this time, the lower leaves dead and the upper ones yellowed. Seed maturity is immaterial in this case.

Special care is needed with seed baths, as linseed will easily go mucilaginous if wetted. Fibre-producing varieties are treated with yarrow, linseed-producing ones with valerian. In this particular case cold rather than warm water is used (2 litres/100 kg, 1 quart/100 lb) and the bath is applied in split doses. Spray the seed first with one third of the required amount, ten minutes later with the second and fifteen minutes later with the third third, gently turning the seed with a shovel between treatments. It is definitely worth the effort to achieve more rapid development and healthier and more resistant plants.

Horn manure is applied when drilling and before or during one pass with the mechanical hoe; horn silica is sprayed when the plants are 25 cm (10 in) high, when buds are fully developed and after flowering.

Valerian is sprayed just before buds begin to form, as with all flowering plants.

Linseed yields are 1.5–2 t/ha (0.65–0.9 tons/acre). The harvest from about 0.1 ha (¼ acre) is sufficient for about 15 calves.

---

* Linseed from crops given heavy applications of mineral nitrogen may produce traces of prussic acid when made into a decoction to feed young stock; only biodynamically or organically grown linseed should therefore be used.

Mustard

White mustard *(Sinapis alba)* is one of the oldest cultivated plants and very much a long-day plant. It therefore needs timely drilling to grow for grain, between the end of March and mid-April, though late frosts may cause damage. Later drilling means that the plants come into flower too soon, before they are fully developed, resulting in extremely low grain yields. It is best to wait with stubble-sowing for green fodder until mid-August or later to ensure a long period of vegetative growth when green matter is produced.

Seed rate is 10–12 kg/ha (9–11 lb/acre) with rows 20–25 cm (8–10 in) apart; denser crops will yield less seed.

Manuring, cultivations and seed bath are as for rape.

Mustard grains from plants grown as a support for peas are easily separated out and provide seed for green manure and fodder crops.

Buckwheat

Buckwheat *(Fagopyrum esculentum)* is attractive to bees and deserves brief mention as there is some demand for the grain.

Grain crops will only do well on light or loamy soils. Buckwheat is fast-growing and frost-sensitive, so that drilling at a seed rate of 60 kg/ha (55 lb/acre) is done in mid-May, and in mild, warm climates also after early-clearing cereals up to the end of July. Rows are spaced at 15–20 cm (6–8 in), drilling depth is 3–2 cm (1 in).

Cultivation is not necessary as the crop soon covers the ground, suppressing all other plants. Manures apt to force growth should not be used if growing for grain as they encourage vegetative growth at the cost of seed production.

The crop is harvested when the majority of seeds show a certain hardness and the colour of ripeness. This is chestnut brown for the smooth-seeded species *(F. esculentum)* and a uniform grey and brown or greyish black for the rough-seeded species *(F. tataricum)*. Harvesting is as for phacelia.

Buckwheat is not grown for fodder because of its fagopyrin

content and low volume of greenstuff. Fagopyrin causes fagopyrism in animals, making them sensitive to light.

> Buckwheat is one of the best bee plants.

Stinging nettle

Generally considered a weed today, the stinging nettle *(Urtica dioica)* was widely used in human and veterinary medicine in the past. Nowadays, fresh plants, nettle juice, the dried herb and alcoholic extracts are used

- in medicated feed,
- for rearing young stock, especially poultry,
- as an additive to dung liquor and in compost production,
- as plant liquor to overcome setbacks in crops and for pest control,
- to produce the biodynamic compost preparation,
- in natural medicine and
- as a health food or vegetable.

Stinging nettle is perennial and its flowers are generally unisexual and on separate plants. Yellow roots, much branched in young plants, less so but thicker in older specimens, give rise to erect pale green or reddish violet square stems. The leaves are in opposite pairs always at a right angle to the preceding pair; they are dark green, hairy, heart-shaped and pointed with toothed margins. In May–October, flowers grow in pseudo-spikes from the upper leaf axils of side shoots.

It would be difficult to find sites and conditions where the stinging nettle does not manage to grow. It also seems to have a particular affinity for human habitations, always growing in the immediate or almost immediate vicinity. Transitional sites are preferred, e.g. the edges of woods, hedgerows, fences, among old buildings and unused machinery, on the banks of streams and ditches, in rubbish dumps and soils rich in nitrogen. These are the places where the plant really thrives.

Boggy and loamy soils rich in nutrients and with adequate moisture are suitable for nettles. Dry sites do not suit.

Nettle seed does not germinate easily and needs below zero temperatures to do so. It is therefore easier to use root sections or mature stolons for vegetative propagation. For relatively large crops, rhizomes may be collected from friable soil in early spring or in autumn using a rotary cultivator or harrow or a cultivator moving at a slow speed. Each is divided into three or four cuttings. Stolons may be up to 60 cm (24 in) in length and should be cut in sections of not less than 15 cm (6 in).

Planting is done by machine or hand, using dibber or trowel, in weed-free friable soil fertilized with manure, partly rotted-down woody material or compost. Space rows to half the width of the tramlines (62.5 or 75 cm, 25 or 30 in), with an inter-row spacing of 50–60 cm (20–24 in), giving *c.* 25,000 plants per hectare (10,000/acre).

Cultivation operations must be done early as stolons will soon sprout. No further cultivation should be done in subsequent years as it may damage the runners that are spreading just below the surface and cause growth to be delayed.

Long-term crops are fertilized with untreated raw liquid manure, as this meets the special need of nettles, which like to change chaotic soil conditions liable to lead to forced growth into a balanced good crumb state. It is not advisable to use fresh farmyard manure containing a high percentage of straw, as unrotted straw will get mixed up with the harvested nettle material and cannot be separated out.

Nettles are highly sensitive to compacted soil and to the pressure from tractor wheels when cultivations and manuring are done on soft, wet ground.

Utilization begins in early spring. Fresh shoots may be eaten as a vegetable and used to make tea and juice for 'spring cures'. Nettles are ready for cutting for medicinal use when the leaves are fully developed. The flowering plant is dried for medicated feed and used to make the biodynamic preparation.

Large areas are harvested with available mowers. A double knife bar set fairly high is best for rapid aftermath and a good quality cut. Rotary windrowers have proved useful for careful windrowing of the plants which have been allowed to wilt only slightly and are 120–180 cm (4–6 ft) long, and for loading them on to trailers. Warm or hot air drying is the only method of preserving relatively large quantities without losses. The field must be level and stone-free to avoid contamination of the product and making it unsaleable.

Depending on site and weather conditions, two or three cuts may be made a year. The nettles lose their vigour after four to six years. Dandelion and white clover, too, may spread. The crop is no longer marketable but still provides good supplementary feed for the farm animals.

The average annual yield is 5 t/ha (2 tons/acre) of dried nettles for sale.

Fodder crops

As stated in the chapter on grassland management, efficiently cultivated and manured meadows and pastures provide the ideal basis for supplying farm animals with a rich and varied diet. Fodder crops are equally important as they

- provide the additional feed required and thus meet protein requirements,
- take over the function of permanent grassland on entirely arable farms,
- suppress and control weeds and
- maintain and improve soil fertility (humus levels).

Legumes meet these requirements particularly well. They provide well-structured roughage, give high protein yields and also increase soil nitrogen levels. Long leys also provide fine grass roots and are the only certain method of increasing the permanent humus level in soils. This offers ideal conditions for commercial growing of row crops, cereals and other crops.

If the soil and the climate are right, fodder can be largely grown as catch crops, so that less of the available arable land is needed to grow fodder as a main crop.

The available range of fodder plants is large, so that suitable species and varieties can be found for all soils and climatic conditions. Lupin and serradella are particularly suited to very light soils, Alsike clover to damp, heavy soils, and birdsfoot trefoil with meadow fescue to relatively dry sites.

> Meadows and pastures provide the ideal basis for supplying farm animals with a rich and varied diet.

In cropping sequences, perennial clover-type crops usually follow cereals, whilst fast-growing annual fodder plants are grown as second crops. Almost all are considered good preceding crops for row crops, cereals and oil plants as they generally increase humus and nitrogen levels, improve soil structure and tend to be deep-rooted.

Within-crop and inter-crop plant compatibilities require attention, especially with sunflowers, flax, red, Alsike and black medick, sainfoin and peas.

Fodder plants are frequently grown as undersown crops and make special demands on cultivation in so far as the seed tends to be small and needs a fine, firm seed bed; this should be as free from weeds as possible as growth tends to be slow in the juvenile stage. In difficult conditions, open drilling (without cover crop) may be the best method, as cultivations can then be adapted to the special needs of the crop.

With regard to soil fertility, all fodder plants, with the exception of serradella and yellow and blue lupin, need dependable supplies of lime and phosphorus. Non-leguminous plants (crucifer family, grasses) usually need high nitrogen levels. Cover crops grown with legumes may need timely applications of manure, if possible well rotted farmyard manure.

Seed quality must be high even for fodder crops. Poor quality results in incomplete, stunted crops, the negative effect being particularly marked in long-lived crops. The field will be left in poor crumb, with nutrient levels low, full of perennial weeds, and yields will be low.

Extensive leys with partly bought-in seed will be increasingly fouled with weeds in later years, mainly curled and broad-leaved dock *(Rumex crispus* and *R. obtusifolius)*. It is easy to see why farmers are increasingly producing part of their own seed from fodder plants.

In the past, the origin of small-seeded fodder plants was most important, for the provenance had to suit the site to give good yields. Propagation for seed would be done in the districts from which the names derive or in sites offering similar conditions. Farmers knew which grew best and gave persistently good crops in which location.

Drilling technique also adds much to the result. Fodder plants are given a good start by suitable cover crops, especially cereals, for stubble of reasonable length provides cover from drying winds, strong sunlight and impact from heavy rain. (For sowing technique, see also pages 39–42.)

Cereals suitable as cover crops are listed below in order of preference (Renius 1978).

Winter crops

1. Rye provides the best conditions as it gives little shade and is harvested early.
2. Wheat is slow to develop in spring and can be harrowed a number of times. This provides a good seed bed and protects the juvenile stages of the undersown crop.
3. Barley tillers in autumn and has therefore produced dense cover by spring. The risk of lodging is also greatest. The advantage lies in the early harvest.
4. Rape ranks about the same as barley.

Spring crops

1. Early maturing wheat varieties are the best cover crop as they are resistant to lodging, let the light through and are drilled early.
2. Barley needs relatively little water and clears early. The tendency to lodge may be a disadvantage. Vigorous clovers may grow through if drilled too early; drilling should be delayed until the barley is at the four-leaf stage. If this coincides with a regular dry period, another cover crop should be chosen.
3. Oats are perfectly suitable as a cover crop if early maturing, even if they do come last in this list. Growth tends to be dense and this may inhibit the undersown crop. On dry sites, the high water consumption of oats may cause problems, so clover and oats should be drilled at the same time in this case.

Red and white clover and Alsike do well if drilled early; they may even be drilled into light snow cover, providing the soil is not frozen. Lucerne (alfalfa), black medick and sainfoin need warmer conditions and should not be drilled before the end of March.

Early clearing crops provide the best conditions for oversown crops, especially a mixture of vetches, peas and oats grown for silage.

The following grasses are particularly suitable for autumn drilling with winter cereals. They are listed in descending order of late drilling tolerance (Renius 1978):

- timothy (Phleum pratense)
- creeping red fescue (Festuca rubra var. genuina)
- meadow fescue (Festuca pratensis)
- cocksfoot (Dactylis glomerata)

9. ARABLE CROPS

crop	seed rate (kg/ha)	sowing time
Persian clover	15–20 (O)	April
+ annual ryegrass	+ 10–15	
Egyptian clover	30 (O)	end of April/mid-May
+ annual ryegrass	+5	
serradella	40	mid-May
red clover	15–20 (20–30)*	spring–August
+ Italian ryegrass	+ 5–25 (15–25)*	
lucerne	15–18	mid-April/mid-August
black medick	20	March–July
Alsike clover	8–10 (3–5)**	spring–August
white clover	10	spring
sainfoin	140–160 (U)	spring
white melilot	20 (U)	spring to early September
O open sown; U undersown; * tetraploid; **as part of mixture		

Table 26. Seed rate and sowing times for clover-type fodder plants grown as main crops.

If timothy is drilled early with winter barley, the plants may start to shoot in spring, resulting in loss of yield and problems with harvesting the cover crop. Creeping red fescue may be grown on all soils for green manuring or mixed with timothy and meadow fescue, which is not very competitive, in permanent leys. Timothy grown in a mixture with fescue or cocksfoot needs to be resown in October. Clovers and other grasses and herbs are drilled in spring at an angle of about 35° to that of the winter crop. Any soil preparation that is needed may be done with weed harrow or roll; rolling after drilling is an advantage.

Drilling depth is low with most clovers and grasses, but drilling is still preferable to broadcasting by hand or machine as it ensures that the seeds are covered with a fine layer of soil and seedlings are protected from drying out.

Ridging must be avoided when using a mechanical hoe. The small elevations in the soil present a serious problem when haying and cutting for silage.

Seed baths also benefit small seeds, but some species tend to go mucilaginous. It is therefore necessary to check the seed repeatedly when applying the liquid, using less or discontinuing the operation in case of doubt. When the seed is spread to dry, it may be necessary to turn it more frequently.

valerian	birch pit concentrate	wild camomile	yarrow
maize	phacelia sunflowers	all legumes marrowstem kale oil radish rape turnip rape mustard turnip	fodder rye grasses

Table 27. Seed baths for fodder plants.

Horn manure applications encourage growth to a marked degree and contribute a great deal to the success of undersown and stubble-sown crops.

When horn silica is applied to the cover crop, this must provide sufficient leaf mass to protect the young clover and grass seedlings from being wetted, as their growth may be inhibited. If this is not possible, the silica preparation should be sprayed at a later date.

Annual clovers

Persian clover (*Trifolium resupinatum*)

This originated in Iran and Afghanistan; it is only in recent years that it has been more widely grown in Central European latitudes. It is not winter hardy, but high quality cultivars tolerate late frosts reasonably well. Persian clover does well in soils that have a good supply of lime, but it prefers the better sites.

Rapid and vigorous growth means that whilst not suitable as an undersown crop with cereals it increases the fodder value of green fodder mixtures. Silage from these has an excellent protein/starch ratio which gives it special value as a feed for dairy cattle.

Persian clover grown on its own and cut after full flower is eaten greedily by the animals and will give high milk yields. Feeding young Persian clover on its own is not advisable because of the risk of scouring and bloat. The solution is to mix the clover with fast growing annual rye-grass *(Lolium multiflorum* var. *westerwoldicum)*. In sufficiently warm and wet weather conditions Persian clover can be cut up to five times.

Applications of horn silica when the crop is hand high and about a week before utilization will reduce the risk of bloat and improve growth. Dense crops cut a number of times will suppress creeping thistle and other persistent weeds within one growing year; they also have a high preceding crop value for row crops and spring cereals.

Egyptian clover (*Trifolium alexandrinum*)

This has similar properties to those of Persian clover. It originates from the Mediterranean region and is sensitive to late frosts. With both species the provenance of the seed is important. Only multi-cut cultivars produced in Europe, mainly Portugal, should be used. Egyptian clover has a lower crude protein and dry matter content than red and Persian clover. Cattle find it palatable, but it is only useful as green fodder.

An early cut encourages aftermath. Weed suppression and preceding crop value are slightly less than for Persian clover.

Serradella (*Ornithopus sativus*)

Serradella will grow in slightly acid, slightly sandy and almost boggy soils if there is sufficient water. It does not like excessive lime or dry conditions. The varieties grown in Central European latitudes are sensitive to temperatures below −6°C (21°F) and therefore not winter hardy. Undersown or stubble-sown serradella produces tender, non-woody green fodder suitable for soilage, silage, grazing and green manuring.

Mixed with Alsike and white clover and undersown, or with yellow lupin, millet or mustard and stubble-sown, it may be expected to give a good yield.

Perennial clovers

Red clover (*Trifolium pratense*)

This is used on its own or with perennial ryegrass mainly over a two year period. Timely undersowing with suitable cover crops will usually still allow one good cut in the first year, with two or three cuts the following year.

Very loose (boggy), light (sandy) and poorly drained soils are not suitable. In the right site, red clover gives high yields for a variety of uses. Grown on its own it is mainly used as green fodder. For silage or hay it has proved useful to mix it with Italian ryegrass *(Lolium multiflorum* var. *italicum)*. It likes to be shallow sown and is therefore better drilled after the cover crop. In areas with dry periods in spring it needs to be drilled as early as possible, best of all in winter rye.

The best time for mowing is before full flower. Good strong roots will help overwintering. Lush foliage by the beginning of winter, on the other hand, results in winter injury; late utilization is therefore important. The crown from which the shoots grow is drawn close to the soil as the tap root contracts in autumn, so that the plant is well protected in winter. If leaves have been left covering the crown, the plant may begin to rot under snow cover, resulting in missing plants and slow development in spring.

Tetraploid red clover mixed with tetraploid ryegrass is very winter hardy and gives excellent yields. Aftermath capacity even in relatively dry sites, weed suppression and good legume tolerance have made this a popular crop. It grows on well into autumn, which makes it a better preceding crop for spring crops or row crops than for winter wheat. Careful ploughing is needed or it will grow through again, being vigorous and winter hardy, and interfere with the following crop.

Lucerne or alfalfa

Lucerne *(Medicago sativa;* hybrid lucerne, a cross of blue-flowered *M. sativa* and yellow-flowered *M. falcata)*, the queen of fodder plants, was called a 'child of the sun' by Rudolf Steiner. In the warm sun of the Mediterranean, in southern France, it allows maximum yields with seven cuts. Protein yields can be very high, twice as high than for red clover, four or five times as high than for oats or a poor meadow.

Ideal conditions are warm, permeable loess, as deep as possible and rich in lime, or warm weathered limestone soils. Lucerne follows a weed-free row crop, either under a cereal cover crop or open drilled.

Yield and the period of utilization depend greatly on the timing and number of cuts. Timing depends on

- the stage reached by the buds and shoots developing on the root collar (rhizome). These will be the aftermath and must not be cut. Care must be taken therefore not to cut too late or too deep;
- storage of carbohydrate reserves in the root, which happens at full flower. The crop should therefore only be cut at this time.

Red clover will give high yields in most sites except those that are very dry in summer.

Final utilization in the year should be such that sufficient assimilates can be stored. This means that the last main fodder cut should be done no later than mid-September.

Spring cultivations are harrowing after drilling in the first year and, after this, grubbing to loosen the top soil, in areas where there is a risk of late frosts after the first cut. In the year when the lucerne is ploughed in, the operation is followed by a drilling of annual ryegrass. The fine roots of this leave a good crumb structure, the lucerne roots having got the middle and subsoil into first class condition. The preceding crop value of lucerne (alfalfa) is enhanced by this method.

Black medick (trefoil) (*Medicago lupulina*)

This is a species of lucerne but is annual or at most biennial. It demands a lot of lime, but even shallow, cold and stony weathered limestone soils will suffice. As a low-growing species, black medick is mainly undersown for green fodder and green manuring.

Alsike clover (*Trifolium hybridum*)

This holds a position midway between red and white clover as regards the colour of its flowers and growth characteristics. It nevertheless ranks as a species in its own right. True hybrids of red and white clover are not possible.

Alsike clover will still manage quite well in cold, wet sites where red clover does not thrive. Only dry, lean or hot soils are not suited. Alsike clover is only grown on its own for seed, as yields are low and bitterness makes it unpalatable unless mixed with other fodder. Given to horses, it may cause severe skin inflammation.

The most important use is in leys. It is self-compatible and not very susceptible to clover stem rot, which is why it is used in place of red clover in sites where this is at risk.

> Grown in suitable sites and efficiently utilized, a crop of lucerne (alfalfa) will give top quality yields for several years.

White clover (*Trifolium repens*)

This clover has multiple uses and is highly adaptable. It is essentially a grazing plant and as such persistent; what is more, frequent mowing and grazing actually enhances development.

White clover grows on practically all sites, from wet bogland to dry sandy soils. It does best on fresh to moist sites, but adequate atmospheric humidity will suffice. Here the large-leaved tall varieties of white clover grown with perennial ryegrass *(Lolium perenne)* give high fodder yields. Trials done in England have shown that protein yield was equally high for perennial ryegrass grown with 30% of white clover and no mineral fertilizer and grown without clover and given 600 kg/ha (525 lb/acre) of nitrogenous fertilizer.

White clover is winter hardy, self-compatible and usually grown for three or four years; aerial runners and self-seeding frequently give it almost unlimited persistence.

Crimson clover (*Trifolium incarnatum*)

Crimson clover is mainly known as part of the Landsberg mixture (see page 190). It is the fastest growing clover species and ready to cut in spring two or three weeks before red clover if grown for more than a year. Winter damage is frequently due to late drilling (after 15 August) and lack of firmness in the soil.

The species does well anywhere except in extremely wet and cold soils, and in boggy and lean sandy soil.

Birdsfoot trefoil (*Lotus corniculatus*)

This is an undemanding persistent legume. It is winter hardy, stands up well to grazing and cutting and tolerates relatively long periods of dryness. It is rarely drilled as a pure crop in Central Europe, probably because local varieties are not suitable for this. On the other hand it is common in pasture and meadow seed mixtures and in ley mixtures for dry sites. Relatively high proportions in a ley ration given to dairy cows will benefit milk fat levels.

Greater birdsfoot trefoil (*Lotus uliginosus*)

This close relative to birdsfoot trefoil prefers really damp sites. It merits more frequent inclusion in permanent leys for damp and wet areas, for it is winter hardy, persistent and tolerant of shade. Grown under the right conditions and in sufficiently high proportion it enhances the normally low fodder value of such leys because of its high protein content.

Birdsfoot trefoil cultivars available in the USA give good fodder yields on heavy soil in dry sites not suited to lucerne (alfalfa).

Sainfoin (Onobrychis viciifolia)

Sainfoin loves warm, dry calcareous soils. Unlike *O. montana* and *O. arenaria*, which demand a lot of lime, sainfoin is no more demanding than lucerne and will also do well in shallow lime, chalky and marly soil. Even sandy soils will do, providing the subsoil contains marl. The species does not grow in wet, cold and impermeable soils nor in almost boggy ground. It needs less warmth but more light than lucerne, especially when young. Undersown sainfoin will therefore do best if drilled between the rows of the cover crop on the light-well principle, i.e. following emergence of the cover crop which is grown in rows spaced at 25 cm (10 in). Sainfoin is sensitive to late frosts in its juvenile stages and is therefore not drilled as an undercrop until after the beginning of April. Drilled as a pure crop after a winter catch crop on suitable sites it will give good results if drilled as late as the end of May and beginning of June.

The crop is extremely sensitive to grazing and cutting in the rain or with rain soon following. The hollow stubble rots easily and the whole crop perishes.

Sainfoin has a strong tap root and almost equally strong lateral roots; these branch widely before they also go down, penetrating fissured rock with a dense network of fibrous roots. The plant takes possession of the soil as it combines the properties of widely branching red clover roots and the more deeply penetrating roots of lucerne (Kraus 1914). The roots are reddish brown to sulfur yellow; the woody root bark with its cork layer is well protected from water loss due to evaporation, so that sainfoin is extremely resistant to drought conditions.

Two types are grown. The single-cut, late-flowering type will give one good cut at the end of full flowering and then remain in the rosette stage. The double-cut type flowers earlier and is cut when it begins to flower. It is highly sensitive to premature or late cuts and must be given time to grow between two utilizations. Otherwise vitality and yield will go down and grasses will quickly take over.

Sainfoin provides excellent, palatable fodder, with higher yields than lucerne on soils where lucerne is near its limit.

The crop will persist longer if about 18 t/ha (8 tons/acre) of composted manure are applied every two years. Sainfoin is not merely useful but also truly handsome as a plant. On poor, shallow and stony calcareous ground, the rose pink field, alive with the hum of bees gathering nectar, is a delight to the farmer's eye. It is one of the best preceding crops for cereals, grasses, roots and tubers and its very presence in arable land will make all cereal crops more fruitful. Sainfoin should therefore be much more widely grown. Yields may be slightly less than with other legumes, but this is made up for by the beneficial, harmonizing effect on the whole farm.

White melilot (*Melilotus alba*)

White melilot is a pioneer plant. It improves sandy, shallow and dry soils that have adequate lime by producing a large volume of dry matter and because of its great capacity for dissolving minerals. The high coumarin content means that it cannot be used for fodder, and it also grows woody quite early. It suppresses weeds completely.

Leys

Well managed meadows and pastures are an ideal most easily achieved on arable land with rich mixtures of clovers, grasses and herbs. Apart from high fodder quality, these are also of great benefit to soil fertility.

Unbalanced mixtures, e.g. red clover and ryegrass, or lucerne and false oatgrass, will give high yields, especially of protein, but they will never equal the fodder value of mixtures rich in species. The latter also leave more root residues, with a more lasting effect on soil structure and humus levels,

- due to specific root activities,
- due to differences in ability to dissolve minerals, and
- due to the differences in mineral composition and concentrations in the plant matter which are the result.

Richly varied and well balanced rations are vitally important for farm animals. It has been shown that dairy cattle given unbalanced rations, e.g. lucerne hay and beet and turnip leaf silage, develop fertility and other health problems that are difficult to overcome. Cows allowed to graze in pastures rich in herbs and fed with hay and silage from meadows and leys containing many species have been found to be more fertile and stable in health. Many grassland herbs contain high levels of trace elements that help to prevent deficiency diseases if available in the right proportions. Selected herbs should therefore be added to fodder crop mixtures.

> Balanced rations with plenty of variety prevent infertility and other health problems in farm animals.

The following points need to be considered when devising clover, grass and hay mixtures:
- soil and climatic conditions, especially water supply during the main period of active growth;
- intended use as pasture, for cutting and grazing, or cutting only;
- intended use as hay, soilage or silage;
- intended period of utilization;
- qualities and requirements of individual fodder plants.

To find out which are the best mixtures for a particular farm, farmers need to have a detailed knowledge of the properties and requirements of different plant species and varieties, and need to make careful observations of wild plants natural to the site.

When a carefully devised mixture has been drilled, development over subsequent years needs to be monitored so that necessary changes may be made to later mixtures, taking account of legume compatibilities. A clover, grass and herb crop should contain about 45% of clover, 45% of grass and 10% of herbs.

With long leys it is important that plenty of good fodder is produced not only in the first but also in later years. Consideration therefore needs to be given to the persistence of the species included and also to their ability to compete against weeds.

Table 28 gives the most important qualities of commonly used grasses.

It is impossible to give standard formulas for suitable mixtures, for every site is different and the number of potential mixtures almost equals the vast variety of sites. Successful drilling therefore depends on an accurate knowledge of local conditions and especially the microclimate. Site adapted plants are chosen that as far as possible are harvest ripe at the same time, mutually compatible, and complementary as regards nutrient requirements, ability to dissolve minerals and root space requirements. They should also provide palatable fodder of rich variety and of a high nutrient value. The choice is difficult as there are so many varieties. With rich mixtures, the components may be expected to be to some extent complementary. Tables 23, 28, 29 and 30 provide some useful information.

Table 30 gives figures for the relative seed quantities required.

The drilling method is that described in the chapter on re-seeding grassland.

The duration of a ley determines the extent to which permanent soil fertility is improved. Two or preferably more years create excellent conditions. It will however be necessary to maintain a dense crop for the whole duration. This depends on

- using the right kind of mixture,
- a well-prepared seed bed (free from perennial weeds) and
- weather conditions.

A period of two or at maximum three years is generally best, as this allows the good preceding crop effect to be brought to bear more quickly and more frequently.

name (botanical name)	habit	life	time of coming into ear: days after April 1
sheep's fescue (*Festuca ovina*)	bottom grass compact tufts	perennial	15–39
meadow foxtail (*Alopecurus pratensis*)	top grass loose tufts	perennial	24–26
creeping or red fescue (*Festuca rubra*)	bottom grass with stolons	perennial	39–49
smooth meadow grass (*Poa pratensis*)	bottom grass with stolons	perennial	37–48
swamp meadow grass (*Poa palustris*)	bottom grass loose tufts	perennial, not long-lived	63
cocksfoot (*Dactylis glomerata*)	top grass compact tufts	perennial	44–59
perennial ryegrass (*Lolium perenne*)	bottom grass	perennial dense tufts	46–71
meadow fescue (*Festuca pratensis*)	top grass	perennial	50–57
Italian ryegrass (*Lolium multiflorum*)	top grass loose tufts	annual or longer	52–61
false oatgrass (*Arrhenatherum elatius*)	top grass	perennial	52–55
yellow oatgrass (*Trisetum flavescens*)	middle grass loose tufts	perennial	52
hybrid ryegrass (*Lolium hybridum*)	middle grass compact tufts	biennial or longer	58–63
timothy (*Phleum pratense*)	top grass	perennial	62–92
tall fescue (*Festuca arundinacea*)	top grass	perennial	59
fiorin (*Agrostis alba*)	middle grass loose tufts	perennial, does not compete well	69–72
annual ryegrass (*Lolium multiflorum* ssp. *gaudini*)	top grass	annual, not winter hardy	71–79

Table 28. Key characteristics of some grasses.

9. ARABLE CROPS

recovery	site requirements: climate	soil	uses
moderate	all sites	pasture also dry	pasture green manuring
good	damp, fresh	not too light	meadow
excellent	all sites, incl. harsh climates	all, incl. wet	pasture
good	all sites	all, except wet, heavy	meadow and pasture
good	damp	damp, wet	meadow and pasture
excellent tolerates dryness	damp	all, incl. light	grazing none too frequent
excellent	high rainfall	all damp soils	pasture, frequent grazing
good	high humidity	medium to better quality	pasture, sensitive to overgrazing
excellent	high humidity high rainfall	almost all	winter catch crop
good	warm, not too wet	well structured, rich in lime	not for grazing, not to be cut too often
good	dry, warm	permeable rich in lime	pasture and meadow
good	high humidity	almost all	winter catch crop leys
good	fresh and damp	medium to good	pasture
moderate	wet and cold	heavy, cohesive	early mow pasture
moderate to good	wet and cold	damp	pasture
excellent	damp	high in nutrients cohesive	alternative to clover

173

a) As main crop		
dry sites not suitable for red clover and lucerne	sites suitable for red clover and lucerne	damp, cool sites, cold soils
black medick birdsfoot trefoil sainfoin cocksfoot Hungarian brome	red clover lucerne Italian ryegrass false oatgrass	Alsike clover birdsfoot trefoil timothy meadow fescue
b) For grazing, perennial ryegrass and white clover must be included. c) Intended period of utilization			
stubble sown	18 months	2½ years	long-term
red clover Alsike clover white clover black medick serradella Westerwold's ryegrass	red clover Alsike clover black medick Italian ryegrass timothy cocksfoot meadow fescue false oatgrass	Alsike clover birdsfoot trefoil (or lucerne) white clover timothy cocksfoot meadow fescue false oatgrass	lucerne birdsfoot trefoil sainfoin white clover timothy cocksfoot meadow fescue false oatgrass

Table 29. Selection of mixtures (Klapp 1951).

		purity %	germinating power %*	competitiveness**	amount of seed kg/ha A	B	C
false oatgrass	*Arrhenatherum elatius*	87	77	I			50
cocksfoot	*Dactylis glomerata*	91	85	I			20
Italian ryegrass	*Lolium multiflorum*	96	90	I			20
perennial ryegrass	*Lolium perenne*	96	90	I			30
meadow foxtail	*Alopecurus pratensis*	72	70	II		29	33
Hungarian brome	*Bromus inermis*	92	85	II		41	47
yellow oatgrass	*Trisetum flavescens*	72	70	II		23	27
florin	*Agrostis alba*	90	87	III	7	9	12
meadow fescue	*Festuca pratensis*	95	90	III	45	60	75
creeping fescue	*Festuca rubra*	92	87	III	25	33	41
reed grass	*Phalaris arundinacea*	93	75	III	15	20	25

timothy	*Phleum pratense*	95	88	III	12	16	20
swamp meadow grass	*Poa palustris*	91	87	III	15	20	25
smooth meadow grass	*Poa pratensis*	91	86	III	15	20	25
hybrid lucerne	*Medicago varia*	95	85	I			20
red clover	*Trifolium pratense*	96	87	I			20
black medick	*Medicago lupulina*	94	82	II		23	27
birdsfoot trefoil	*Lotus corniculatus*	95	81	III	20	26	33
large birdsfoot trefoil	*Lotus uliginosus*	94	81	III	10	13	17
Alsike clover	*Trifolium hybridum*	95	87	III	15	20	25
white clover	*Trifolium repens*	95	87	III	12	16	20

* Good averages
** I = tends to displace components rated II and III
   II = will displace or be displaced, depending on circumstances
   III = at risk from more vigorous components
A In mixtures containing group III species only.
B In mixtures containing group II and III species only.
C In mixtures containing group I–III species.

Table 30. Seed list for mixtures (Klapp 1951)

Should conditions prove unfavourable so that a newly drilled crop fails, it is better to plough the field and drill a mixture of short-lived fodder plants such as Egyptian and Persian clover. The same applies if bare patches appear early or the field is fouled with weeds.

The best method of manuring long-term fodder crops is to apply a good quantity of well-rotted composted manure to the preceding crop or prior to drilling in late summer. Repeat in the second year of utilization. Depending on the type of mixture used, an application of dung liquor would benefit a mixture with a high proportion of grass. In clover areas it is better used on other crops. An application of composted manure and dung liquor before ploughing will enhance the long-term preceding crop effect.

In his book *The Clifton Park System*, which was first published at the end of the nineteenth century, Robert H. Elliot described the effect of adding herbs to the mixture. He worked with a rotation of four years of arable alternating with four years of grassland. A special feature of the system was the addition of five herbs with good root development to the ley mixture:

- chicory *(Cichorium intybus)*
- salad burnet *(Sanguisorba minor)*
- yarrow (*Achillea millefolium*)
- parsley *(Petroselinum crispum)*
- wild parsnip *(Pastinaca sativa)*

The complete mixture was as follows (kg/ha and lb/acre):

cocksfoot	12.5	11.0
tall fescue	6.0	5.5
oatgrass	6.0	5.5
hard fescue	2.5	2.2
dogstail	1.2	1.1
yellow oatgrass	1.2	1.1
late-flowering red clover	2.5	2.2
white clover	2.5	2.2
Alsike	2.5	2.2
kidney vetch	4.0	3.5
chicory	5.0	4.5
burnet	10.0	9.0
yarrow	1.2	1.1
parsley	1.2	1.1
wild parsnip	1.2	1.1

Having gone through the rotation twice, Elliot was able to grow arable crops (twice each row crop and cereal in the four years) without the need for mineral fertilizers, and achieved excellent yields. The lesson to be learned from this is that a rich and varied mixture of site-adapted plants, including suitable herbs, makes all the difference to the soil. Again, duration is a factor, one-year leys being considered unsuitable.

The system has been used at Talhof Farm since 1951, suitably modified for poor, shallow weathered Upper Jurassic limestone and an extremely harsh climate. In over thirty years, the humus level of the arable soil has risen from just below 1% to 4.5%. The resulting improvement in soil structure has increased the apparent depth of working from 4–12 cm (1½–5 in) to 6–24 cm (2½–10 in). The soil is clayey and heavy, but it is possible to use a three-share plough coupled to a 50 hp tractor to a depth of 24 cm (10 in), moving at a good pace.

Cutting needs to take account of the type of mixture and climatic conditions. The same seed mixture will produce growth of different composition on different soils, even within the same year. Dry weather, for example, will encourage lucerne, birdsfoot trefoil, cocksfoot and oat-grass, wetter weather red, Alsike and white clover and meadow foxtail.

Cutting times will depend on the maturity of the constituents. Mixtures with a high proportion of lucerne and sainfoin at the juvenile stage will not tolerate cutting back, whereas open-sown Persian, Egyptian and red clover can be cut back if fouled with weeds. They will grow all the better afterwards, as they have more light, air and warmth and there is no competition for the available water.

Leys should never be cut too short, as some species are sensitive to this. The resting period before the final autumn cut should be long enough to allow sufficient carbohydrate to be stored in the roots.

The first year after drilling the first cut should not be too early to avoid suppression of slow-growing plants.

As to cultivations, very loose puffy soils will need rolling. More thorough work with harrow or grubber should be limited to pure lucerne (alfalfa). A planker is used after compost applications and to deal with mole and vole hills.

The biodynamic sprays are used as for grassland.

For two or three-year leys	
red clover	1
lucerne	4
white clover	2
sainfoin	10
black medick	2
birdsfoot trefoil	8
kidney vetch	1
sweet vernal grass	2
cocksfoot	2
timothy	4
false oatgrass	2
meadow fescue	2
Italian ryegrass	2
perennial ryegrass	2
common meadow grass	2
creeping fescue	1
yellow oatgrass	2
caraway	1
lesser burnet	8
Long-term leys	
red clover	1
lucerne	2
white clover	4
sainfoin	20
birdsfoot trefoil	10
kidney vetch	1
cocksfoot	2
timothy	4
false oatgrass*	4
meadow fescue	2
perennial ryegrass	4
creeping fescue	1
yellow oatgrass	2
caraway	1
lesser burnet	8
chicory	3
parsley	1

* omitted if for grazing

Soil conditions vary enormously (soil numbers ranging from 16 to 59) and the mixture is always specially adapted to the condition and location of the field (slopes facing north, south, east and west).

Table 31. Ley mixtures used at Talhof Farm (kg/ha).

Long leys need to be ploughed up in early autumn 2½ or 3½ years after drilling. This is the best time as regards weed suppression, resting the soil and accumulation of nitrogen and root mass. Both nitrogen levels and root mass are highest in autumn. Ploughing offers the opportunity of deepening the crumb layer. Final utilization is immediately followed by an application of farmyard manure. The land is then ploughed deeply, using a skim coulter. The beams should be at as much of an angle as possible so that sufficient warmth and air can enter the soil and chemical conversions are accelerated.

A long ley with clovers, grasses and herbs contributes a great deal to the continual improvement of soil structure and fertility.

Non-leguminous fodder crops

Non-leguminous fodder crops are not much grown as main crops, unless roots grown for fodder are included in the category.

Mixtures of pulses and cereals may be grown twice in succession in a vegetation period to control weeds. Some fodder plants also give high yields if grown as a second crop on some sites – above all maize for silage, sunflowers, swedes and marrowstem kale.

Maize (*Zea mays*)

Maize growing has seen an explosive increase in Europe since the 1970s. Mechanization is easy, productivity in relation to labour is high, ensilage is straightforward, the material is specially suited to fattening beef bulls, dry matter yields are high and selective herbicides can be used.

A sensible amount of maize in a cropping sequence helps to cover feed requirements and make rations more varied. The proportion should not appreciably exceed 10% of the arable area and 15% of the dry matter in feeds.

On heavy soils it is best to apply rotted farmyard manure or well-fermented liquid manure to the catch crop. No time will then be lost with manuring after harvesting that crop and the soil is not exposed to pressure. On lighter soils maize can also do with manuring in spring. An application of well-humified compost after drilling the maize will give better growth, with the soil warming up more quickly. The compost also encourages root development so that juvenile growth will be more rapid. An additional application of dung liquor will stimulate growth further.

Weed control is easier and the danger of birds eating the grain is reduced by using a ridging tool immediately after drilling. The seeds are then at a depth of 10–14 cm (4–6 in) and difficult for rooks and pigeons to reach. Several brisk passes with the weed harrow will deal with germinating weeds, letting the young maize plants develop undisturbed. If potatoes are also grown it is advisable to use the same row width for both crops, so that implements only have to be set once. If this method is used, cultivation is as for potatoes; for level sowings it is as for beet. Flame weeding has also proved effective.

> Maize yields are definitely increased by repeated applications of horn silica.

Before hoeing the crop for the last time when it is 40–60 cm (16–24 in) high, white clover or black medick may be sown by hand or with the fiddle. If successful, the undersown crop will cover the ground densely by the time the maize forage harvester and trailer are used, greatly reducing damage especially in wet weather. Successfully undersown maize makes a good preceding crop, leaving a well structured crumb rich in root material.

Maize reacts well to repeated applications of horn silica; yields may be increased to a remarkable extent.

Sunflower (*Helianthus annuus*)

This is less demanding in all respects than maize is. The growing period is shorter, weeds are quickly suppressed, fodder yield is high and the soil is left in a good crumb state.

Sunflowers respond well to large amounts of farmyard manure. Seeds only ripen in exceptionally warm areas in our latitudes. The crop is therefore grown mainly for green fodder and silage.

Winter catch crops make excellent preceding crops, sunflowers being sensitive to frost, so that they cannot be drilled until just before or after the last spring frost in mid May. They may be stubble sown after winter rape, winter barley and early-maturing winter rye, peas and spring barley.

Densely drilled crops have thinner stems and this increases the fodder intake of animals. It has proved a good idea to mix a small-grain field pea in with the seed. If cut when the sunflowers are in full bud, or about a quarter are in flower, this mixture is more palatable to cattle than pure sunflower; it can also be used for silage.

The first application of horn silica is made just before the leaves close up the rows.

crop	seed rate (kg/ha)	depth (cm)	distance between rows (cm)
maize for silage	18–52	5–7	62.5–85
sunflowers (second crop)	25–30	3–4	18–25
+ fodder peas	+50–60		
marrowstem kale	4*	1–2	30–40
* for planting, 800 g			

Table 32. Seed rate, sowing depth and distance between rows for non-leguminous fodder plants grown as main crops.

Marrowstem kale

Marrowstem kale *(Brassica oleracea* convar. *acephala* var. *medullosa)* will provide green fodder right until winter sets in, for it will tolerate temperatures down to −10°C (14°F). The green fodder period can thus be extended by several weeks. It will also provide soilage for two or three more weeks if bundles of it are placed in the gateways of barns or along the outside walls of buildings and protected with plastic sheeting and bales of straw. Marrowstem kale is added to maize during ensilage to balance the protein content. Farmyard manure applied to the preceding catch crop or before planting or drilling must be well rotted down, otherwise pests may become more of a problem. About 800 g of seed will be enough to grow seedlings for the optimum number of 90,000 or 100,000 plants per hectare (11 oz/acre for 40,000 plants). Drill in beds in April, six to eight weeks before planting out; this will provide vigorous seedlings for planting as soon as the preceding crop has been harvested.

Cultivations are as for swedes until the rows have closed up.

The first application of horn silica is made as soon as this has happened.

Italian ryegrass (*Lolium multiflorum*)

This will rapidly produce a large bulk of herbage, but requires high nitrogen levels if grown on its own. This is only done in exceptional cases – to fill gaps in clover fields due to winter damage, or on farms with a high percentage of stock where good use has to be found for the resulting manure. Its main use is in combination with red clover in leys of one year or eighteen months. In longer leys it will displace most other species and leave large gaps in subsequent years, since it often does not survive a second winter. Italian ryegrass prefers fresh, light or cohesive soils and a climate that is not too harsh.

Annual ryegrass

Annual ryegrass *(Lolium multiflorum* var. *westerwoldicum)* grows more rapidly than Italian ryegrass, producing mature seed in the year it is drilled. Always open sown and usually also on its own.

Hybrid ryegrass (*Lolium hybridum*)

This is a cross of Italian and perennial ryegrass. It is more long-lived and winter hardy, which makes it suitable for leys that are both cut and grazed.

Catch crops

In sites where climatic conditions permit, catch crops can play an important role in producing additional fodder and green manures. Adequate water supplies are necessary if they are to succeed.

Distinction is made between spring and winter catch crops. Spring crops in particular have been grown more and more frequently in recent years to balance essentially cereal-based cropping sequences. The majority of catch crops belong to the legume and crucifer families. Plant breeding has produced a range of fast-growing varieties.

In borderline sites where harvest is late and active growth ceases early in autumn, undersowing is the only method. This may include all clovers that remain low until the cover crop is harvested, and also leys that will develop quickly once the main crop has been cleared: serradella, white clover, black medick, Alsike, white melilot and late varieties of red clover.

Serradella and lupins, too, may still be broadcast in winter rye as late as May, if rain is expected and the soil is suitable. This is done by hand or, if tramlines exist, with a spinning fertilizer distributor set high, at a time when the rye is 60–90 cm (2–3 ft) high.

The principles are the same as for main fodder crops: the field must be free from perennial weeds, a good water supply is required, and the growth and harvest of the cover crop should not be impeded.

Stubble sowing is preferable in areas where the period of active growth is sufficiently long – if the main crop is harvested early and a long, mild autumn follows – or if perennial weeds are present. Preparing the seed bed after harvest does of course mean some degree of water loss through evaporation, and extensive stubble-sown catch crops do need adequate machine availability.

Every July day gained by drilling early is equivalent to eight August days or the whole month of September as regards the production of leaf mass.

Stubble-sown crops grown to produce high-protein fodder and increase soil nitrogen levels include field peas, vetches and tares, serradella and field beans. If the preceding crops clear early, Persian and Egyptian clover and lupins may be considered singly or in mixtures.

If plenty of manure is available, sunflowers, green maize, marrow-stem kale, turnips, millet, buck-wheat, rape, oil radish, mustard and phacelia will also give good yields.

Winter catch crops are most certain to give good results in the majority of sites if stubble-sown in summer. Different species and varieties have different requirements concerning drilling dates. Rape, Landsberg mixture, vetches and rye, fodder rye and turnip rape (in order of their drilling dates) are most widely grown. Mixtures of Italian, or even better, hybrid ryegrass with rape, crimson clover and winter vetch are generally successful.

A special advantage of winter catch crops is that they prevent leaching of minerals, especially nitrogen, during the cold months and provide early green fodder.

In the following pages, a brief description is given of plants other than those already described that make suitable catch crops.

Turnips (*Brassica rapa var. rapa*)

Grown on lighter soils where water supplies are adequate, turnips give high yields of palatable, milk-producing fodder that may also be used for silage. They will tolerate temperatures as low as −8°C (18°F) and produce sufficient additional growth in autumn to give yields of up to 60 t/ha (27 tons/acre).

All early clearing crops may be grown as preceding crops: green oats and peas in mixture, chitted early potatoes, winter and spring barley, winter rye and early-maturing peas. Cabbages and their relatives should be avoided.

Quick, clean preparation and drilling, shallow working of the soil and rolling after drilling create the necessary conditions for a good yield. Mechanical hoeing is also of benefit. Combines are now available for harvesting.

Afternoon horn manure applications are made when drilling and again when hoeing; horn silica is first applied when the turnips begin to swell, in the morning, with a second application made when the rows have closed up about three weeks before harvesting, this time in the afternoon.

If the following crop requires well manured soil, it is advisable to apply about 25 t/ha (11 tons/acre) of good quality composted manure before or after drilling the turnips.

Liquid manure applied when drilling or later between rows before hoeing will directly increase yields.

crop	seed rate (kg/ha)	depth (cm)	distance between rows (cm)
turnips	1.5–2.5	1–2	25–30
oil radish	20–25	up to 2.5	25–30
Phacelia	8–12	up to 1.5	15–20
summer rape	7–12	1–2	20–30
winter rape	7–10	1–2	20–30
winter turnip rape	9–12	1–2	20–30
Mustard	20–25	0.5–2	15–25
green rye	170–200	2	15–20
vetches and rye:		2–4	15–25
rye	40–80		
+ winter vetches or	+40–70 or		
+ Hungarian vetch	+60–90		
Landsberg mixture:		1–3	15–25
winter vetches	25–35		
+ crimson clover	+15–25		
+ Italian or hybrid ryegrass	+10–20		
(+ winter rape*	+ 4)		

* Winter rape to replace crimson clover in cases of clover sickness

Table 33. Seed rate, depth and distance between rows for some catch crops.

Oil radish (*Raphanus sativus var. oleiformis*)

This has been grown more frequently in recent years, mainly in sugar beet growing areas, for it is inimical to beet cyst nematodes. It is one of the oldest oil plants of Asia but is only used for fodder and green manure in Central Europe.

Being a long-day plant, oil radish will tolerate late drilling, right into the first ten days of September (see Figure 26). It is fast growing and suppresses germinating weeds, whilst its stout tap root penetrates deep even in compacted soils. The mustard oil content rises with age, so that oil radish must be used for fodder or silage by the time the

9. ARABLE CROPS

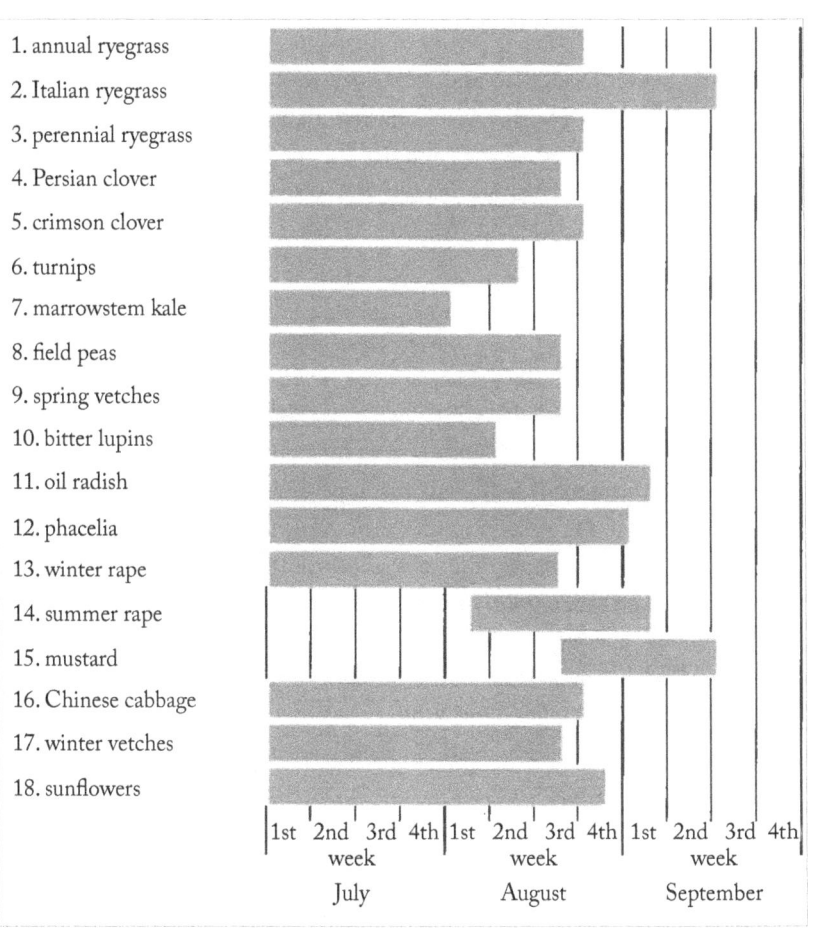

1. annual ryegrass	
2. Italian ryegrass	
3. perennial ryegrass	
4. Persian clover	
5. crimson clover	
6. turnips	
7. marrowstem kale	
8. field peas	
9. spring vetches	
10. bitter lupins	
11. oil radish	
12. phacelia	
13. winter rape	
14. summer rape	
15. mustard	
16. Chinese cabbage	
17. winter vetches	
18. sunflowers	

Figure 26. Favourable sowing times for important catch crops.

pods begin to form. Plentiful applications of liquid manure or dung liquor encourage the production of green mass.

Phacelia (*Phacelia tanacetifolia*)

Phacelia is a catch crop that will quickly produce large amounts of greenstuff; it makes no special demands concerning soil and climate. The only soils that do not suit are those that are very clayey, wet and inclined to form crusts. If plenty of nitrogen is available, a good yield of protein-rich fodder is obtained. Cattle will gradually come to like it if it is introduced with some skill. The plant tolerates temperatures as low as −8° and −9°C (18° and 16°F), but is not winter hardy. It makes a good green manure, leaving the soil rich in root matter, with a good crumb structure and free from weeds. Drilling is possible until early in September.

Summer rape (*Brassica napus* var. *silvestris*)

Summer rape is widely grown as a stubble-sown summer crop. Drilled between early August and early September it will quickly produce high yields if sufficient nitrogen is available. Summer rape may be grown for soilage, grazing or silage. For green manuring, drilling may be at a later date, as a good root mass is produced even if the aerial parts are poorly developed.

Summer rape needs a fine seed bed and is drilled on the flat.

In areas where erucic acid-free varieties of rape are grown, summer rape should be of the same kind to avoid cross fertilization with varieties containing high levels of erucic acid and glucosinolates if the crops reach the flowering stage.

With members of the crucifer family (oil radish, rape, mustard and turnip) grown as catch crops, care must be taken that in the cropping sequence they do not come too close to other members of the family (cabbage, swedes, rape, mustard) grown as main crops; they share the same pests.

Winter rape

This is grown mainly for oil but to some extent also as a winter catch crop. The method is the same in either case, except that earlier drilling is an advantage with the catch crop. This will yield relatively early soilage in spring and is a good preceding crop for potatoes and maize.

Winter turnip rape (*Brassica rapa* var. *silvestris*)

Even earlier green fodder is provided by winter turnip rape. This also suits relatively harsh climates and poor sites. It is drilled three or four weeks after rape. Manuring, drilling method and cultivation is as for rape.

Mustard (*Sinapis alba*)

Mustard will grow rapidly if stubble-sown late and provide soilage or green manure. For soilage, the crop must all be cut before flowering begins, as it then grows bitter and becomes unpalatable. The mustard oil may also affect the taste of the milk. Yields are enhanced by generous applications of dung liquor or liquid manure and by legumes as a preceding crop. Mustard is killed by winter frosts, so there is no risk of it coming up again with the following crop.

Fodder or green rye (*Secale cereale*)

Fodder is one of the most adaptable crops and will grow in almost any soil, the exceptions being extremely poor, dry sandy soils and wet and heavy clays. Drilling should be timely to allow for vigorous tillering in autumn. Liquid manure applied when drilling or on thawing snow at the end of winter will accelerate spring growth and give higher yields. Green rye will provide fodder after green rape.

Digestibility will decrease rapidly once the tips of the awns are appearing; harvest should be complete by this time, therefore, despite the fact that considerable leaf mass is produced later.

Vetches and rye

These need the same kind of soil and climate as pure green rye, but the fodder mixture has significantly higher protein levels. Drilling should be done two or three weeks earlier to allow enough time for winter vetches (*Vicia villosa*, hairy vetch; in warmer regions *V. pannonica*, Hungarian vetch) to develop sufficiently in autumn to be able to overwinter. Higher proportions of vetch will increase the protein level and the preceding crop value of the mixture, but the seed will be more expensive.

> Fodder rye is one of the best preceding crops for carrots.

Landsberg mixture

This has the highest protein and bulk yield of all winter catch crops but also needs the longest period of active growth and a good supply of water. It consists of winter vetches, crimson clover and Italian, or preferably hybrid, ryegrass.

The relative proportions may be varied depending on soil, climate and drilling times. The total seed rate should not be less than 60 kg/ha (55 lb/acre). Drilling should be done by August 20 at the latest to avoid winter damage. The crop needs a fine seed bed prepared sufficiently early to be good and firm. It is recommended to apply composted manure before drilling, without working it in. The amount will depend on the needs of the following crop. Maize for silage, sunflowers, beet grown from transplanted seedlings and marrowstem kale require more than legume fodder mixtures. If annual weeds are likely to be a problem, the distance between rows should be increased from 20 cm to 25 cm (8 to 10 in) so that mechanical hoeing is possible. The slight ridging effect is an advantage.

Lush autumn growth may be cut back or grazed with some care. This will improve the winter hardiness of crimson clover and winter vetches.

Depending on site, the best time for harvesting is between mid-May and early June.

Landsberg mixture may be used for soilage, silage or hay. A dense, vigorous stand will leave the soil in excellent crumb structure and weed free. This reduces the need for cultivation with following crops.

Crimson clover increases the risk of clover sickness in the soil and should be replaced by rape in cropping sequences with a high proportion of red clover and legumes.

9. ARABLE CROPS

	Period of active growth required after main crop harvest	winter rape	winter barley	spring barley	winter rye	winter wheat	spring wheat	oats	early potatoes	mid-early potatoes	peas grown to ripeness	field beans
						preceding crops						
days available for growing catch crops		100	90	80	80	68	59	66	80–100	30–50	90	50
undersown												
red clover	80–100	(+)	+	+	+	(+)	(+)	(+)				
Alsike clover	80–100	(+)	+	+	+	(+)	(+)	(+)				
white clover	80–100	(+)	+	+	+	(+)	(+)	(+)				
clover/grass ley	80–100	(+)	+	+	+	(+)	(+)	(+)				
Italian ryegrass	80–100	(+)	+	+	+	(+)	(+)	(+)				
serradella	80–100		(+)	+	(+)	(+)	(+)	(+)				
fodder carrots	80–100		+	+	+							
Persian clover	80–100		+	+	(+)							
stubble-sown												
fodder kale	80–90	+	+	+	+							
millet	85–90	+	+						+		+	
green maize	80–90	+	+						+		+	
mixed legumes	75–80	+	+	+	+				+			
sweet lupins	70–80	+	+	+	+				+			
turnips	70–80		+	+	(+)				+			
sunflowers	70–75	+		+	+	(+)			+		+	
oil radish	60–75		+	+	+	(+)	+		+			
phacelia	55–65	+		+	+	+	+	+		+		
summer rape	55–65			+	+	+	+	+		+		
mustard	50–60			+	+	+	+	+		+	+	
winter turnip rape	50–70			+	+	+	+	+		+		
winter rape	60–70			+	+	+	+	+				
green rye	45–60			+	+	+	+	+		+	+	+
vetches and rye	50–60			+	+	+	+	+		(+)	+	+
Landsberg mixture	55–70			+	+	+	+	+		(+)	+	
+ well suited; (+) suited within limits												

Table 34. Catch crops within cropping sequences (Fischbeck, Heyland & Knauer 1982).

Growing fodder plants for seed

As previously mentioned, growing one's own seed makes an important contribution to the farm. Fodder crops hold a central position as the link between animal husbandry and arable farming on the one hand and as the mainstay of cropping sequences on the other. A brief discussion of seed production for some fodder plants therefore follows.

Lucerne or alfalfa

Lucerne is grown in a narrow strip between two low-growing crops. Good access for light and air encourages flowering and gives higher seed yields. Two passes with the mechanical hoe, the first with protectors, will encourage seedling growth and suppress weeds. Any weeds that remain and any clover species need to be removed by hand. The crop should not be cut back. Seed is harvested in the year of drilling to keep down flower gall midges. These crops usually give maximum yields the following year because of the carbohydrates stored in the roots.

Red clover

Red clover needs to be grown in dense stands free from weeds. If seed is grown for the farm's use, the presence of other clover species may be acceptable. An early first cut will encourage early flowering with the second growth. The leaf mass will be reduced but flower-heads are larger and more numerous with this method. Dry, warm weather helps flowering and pollination by insects. The crop is mown when the flowers are a blackish brown. The combine may be used in dry, warm years, but in most cases it is best to cut and rack, threshing at a later time.

Black medick

This may be undersown together with perennial ryegrass. The crop is lightly cut for fodder or grazed in the early autumn of the drilling year. Black medick and perennial ryegrass are well matched for growth and maturity, so that they may be harvested together. The seeds are easily separated. After the seed harvest the crop may be grazed.

Alsike clover

This clover has a tendency to lodge, so that relatively dry sites must be chosen for seed growing. Seed is taken from the first cut, which means the year the Alsike clover was drilled if it was undersown into early clearing green mixtures. Otherwise conditions are the same as for red clover.

White clover

Being shade-tolerant, white clover may be grown with suitable cover crops even for seed. The large-leaved tall varieties are taken for seed from the second growth, small-leaved creeping varieties as a first cut. It has been found useful to graze the stand briefly as it begins to come into flower, as the flowering period is long and seed from earlier flowers is simply lost.

Crimson clover or trifolium

Crimson clover cannot be grown for seed at altitudes greater than 400–500 metres (1,300–1,600 ft). Drilling in late summer is preferable to spring drilling. The crop is normally drilled pure and without a cover crop, but may be mixed with early maturing Italian ryegrass. The seeds can be harvested together and it may be possible to cut once more for soilage.

> Fodder crops are the link between animal husbandry and arable farming.

Birdsfoot trefoil

Grasses are included in mixtures of birdsfoot trefoil, meadow fescue, false oatgrass or cocksfoot, or of large birdsfoot trefoil and marsh meadow grass to serve as supports, as both birdsfoot trefoils are unable to stand on their own. With seed grown for farm use it is sufficient to cut the birdsfoot trefoil when wet with dew once the seeds are ripe and spread it on a tarpaulin in a sunny place. Most of the pods should be yellow or yellowish-brown by this time, the grain should have brown cheeks and just break on the nail when tested. If the harvest is kept protected from dew and rain, the pods will split of their own accord when sufficiently dry.

White melilot

This tends to be more of a problem, as many seeds are lost on harvest and seedlings growing from them may seriously interfere with following crops. Seed is taken in the active growing period that follows the year of drilling.

Serradella

It has proved useful to add mustard or summer turnip rape to serradella. The young plants are slow to develop, so that a pass with the mechnical hoe is indicated. If suitable sites are chosen, the plants will soon provide their own dense ground cover. Black medick is harvested when the lower pods have matured and again in the dew, as losses due to breaking may be considerable otherwise.

Italian ryegrass

With Italian ryegrass, seed can only be taken in the second year. Seed may be taken from the first cut in about early July and the second time in early September.

Oil radish

Oil radish for seed is drilled in the second half of April. Poor, acid or extremely dry soils are not suitable. Cool weather at the juvenile stage will improve readiness to flower; later on it is better to have warm, dry weather. A single pass with the mechanical hoe is usually sufficient; after this the plants will suppress any weeds that come up. Composted manure is preferable to liquid manures which are inclined to force growth and favour blossom rape beetle and rape flea attack.

The pods will not burst but are liable to break off, so that seed has to be harvested before dead ripe. Birds may cause major losses when drying in the swathe or on stacks.

Phacelia

Phacelia seed is expensive to buy and therefore best grown at home. One tenth hectare will yield about 35 kg of seed, enough for 3 hectares (¼ acre will yield 75 lb for 7½ acres). Harvest when wet with dew to prevent loss of seed capsules. The time for harvesting is when the last, uppermost flower opens. Drying and final ripening will be rapid with the crop loosely hung on racks. Relatively large areas may be combine harvested, otherwise the crop is thrashed with a flail.

Winter vetches

On farms where winter vetches are grown in quantity, it may be advisable to grow for seed by drilling with winter rye. Wheat would be more resistant to lodging and also has a more convenient harvesting date, but generally cannot be used as it is difficult to separate the seeds. The only possible combination would be wheat varieties with large, long grains and hairy vetch with its small seeds. The seeds have hard seed coats and this may cause problems in so far as dropped seed may persist in the soil and germinate later to create a weed problem. Plant breeders are looking for ways of overcoming this.

crop	seed rate kg/ha	row spacing cm	yield t/10 ha
lucerne	7–9	20–25	2.5
red clover	13–16 (U) 10–12 (O)*1	20–25	2–3
black medick	16–20*2	15–20	3–4.5
Alsike clover	8–10	20	2–3
white clover	9–10 (–12)	25–30	2
crimson clover	25 (–30)	20–25	6–8
birdsfoot trefoil	14–15 (U)*3 12 (O)	20–25	3
large birdsfoot trefoil	12 (U) 9–10 (O)	20–25	3
sainfoin	90–110 (U)*4 70–80 (O)*4	25–35	4–8
white melilot	20–25 (U) 18–20 (O)	20–25	4–5
serradella*5	30–40	20–25	6–7
Italian ryegrass	22–25 (U) 20–22 (O)	20	6–7
oil radish	14–18	20–30	1–1.5
phacelia	8–10	15–20	3–6
hairy vetch	20–40*6	–25	5–8

U undersown; O open sown
*1 If open sown in summer: 12–14 kg/ha
*2 Or 9–10 kg black medick + 15 kg perennial ryegrass
*3 Plus 20–40% of grass
*4 One third less if shelled
*5 Add very small amounts of mustard, summer rape or spring rye to provide support
*6 In mixture with 60–100 kg of rye: sown pure, 90–120 kg (unusual) or 40–60 kg of hairy vetch and 3–6 kg of winter rape in rows 30–40 cm apart. Hungarian vetch – use 20–25% more seed.

Table 35. Seed rate, row spacing and average yields from fodder plants grown for seed.

9. ARABLE CROPS

Herbs

Many plants have a long tradition as medicinal and culinary herbs. In recent years there has been growing awareness of the fact that the healing powers of nature, especially in the form of medicinal plants, can relieve or cure many diseases in both humans and animals. Many native herbs and foreign spices have also returned to favour in cookery.

Herbs can be extremely helpful to domestic animals:
- added to feeds to aid appetite and digestion, to improve hormone balance and, closely related to this, to stimulate lactation;
- as teas, compresses or in ointments to prevent and treat pathological conditions.

In the garden, herbs contribute to the vitality of vegetable crops and help to increase the number of butterflies.

The biodynamic preparations make full use of the special powers of certain plants to produce compost and manure that are able to bring healing qualities, vitality and long-term fertility to the soil.

It is part of the work that is done to make each farm an individual and independent organism to grow one's own herbs and the plants needed to make the preparations. Yarrow, dandelion, oak, valerian, nettle and often also horsetail can be found in fields and woodlands. Camomile (scented mayweed) is easy to grow if not found wild. The method is to collect a basketful of seedlings from a friend's farm or garden and plant these 8 x 20 cm (3 x 8 in) apart in a garden bed. Flowers produced in an area of 10–15 m^2 (110–160 sq ft) will be enough to produce tea and preparation for a 25 ha (60 acre) farm and its animals. Sufficient seed drops out during harvest to provide seedlings on the path beside the bed that can be grown on the next year.

A number of other herbs are also useful to humans and animals. Most culinary and medicinal herbs belong to the carrot family (Umbelliferae), mint family (Labiatae) and daisy family (Compositae). Table 36 gives information on how to grow and use some of the most important ones.

Table 36. Basic data for growing and use of herbs.

name (botanical name)	growing method	space required cm (row spacing)	sowing rate g/100 m² (TGW)	time
carrot family				
angelica (*Angelica archangelica*)	SO p	40 × 50 (30)	15–30 (2–3)	A
anise (*Pimpinella anisum*)	SO a	20–25	200–250	Sp
caraway (*Carum carvi*)	SO p	30 × 30 (30–35)	80–100	Sp/S
celery/celeriac (*Apium graveolens*)	P b	25 × 30 (0.5)	2–3	Sp
chervil (*Anthriscus cerefolium*)	SO a	(10–15)	100	Sp
coriander (*Coriandrum sativum*)	SO a	35 × 40 (25–30)	2000–2500	Sp
dill (*Anethum graveolens* var. *hortorum*)	SO a	20 × 25 (25–35)	80–120	Sp/S
fennel (*Foeniculum vulgare*)	SO p	1st yr: (20–25) 2nd yr: 50 × 60	(4–6.5)	Sp
lovage (*Levisticum officinale*)	SO/D p	40 × 50 (3–3.5)	20	(Sp) A
masterwort (*Peucedanum ostruthium*)	D + St	40 × 40		
parsley – leaf – root (*Petroselinum crispum*)	SO a	(10–15) (25–30)	80–100 40–50	Sp
mint family				
basil (*Ocimum basilicum*)	P a	20 × 20 (20–30)	5–7	Sp
hyssop (*Hyssopus officinalis*)	P/T/D p	30 × 30	6	
lavender (*Lavandula latifolia*, *L. angustifolia*)	P(T) p	30 × 40 (1–1.5)	3–5	Sp
lemon balm (*Melissa officinalis*)	P/T/D p	30 × 40	4–5	Sp/S

p = perennial
a = annual
b = biennial
SO = sown in open
P = plant out
St = stolons

9. ARABLE CROPS

comments	part used	use
tolerates semishade	root herb	pharmaceutics fodder
needs warm climate, germinates in the dark	seed, fruit herb	lactagogue, culinary use, tea fodder
germinates in the light, bee plant, risk of voles	fruit dried stems	culinary use and fodder fodder
sowing not before March, germinates in the light	tuber herb	vegetable fodder (dried late)
	herb	culinary use, fodder
bee plant	fruit straw	culinary use, fodder fodder
bee plant	seeds herb	culinary use lactagogue, culinary use, fodder
divide roots in 2nd year	seeds whole plant	pharmaceutics, tea lactagogue, fodder
(needs singling) plant out	root herb	helps digestion fodder, culinary use
wild plant for hedges, waste ground	root leaf	pharmaceutics, spec. healing properties culinary use, fodder
	herb root	diuretic, vegetable colic, fresh eases pain, culinary use
sensitive to frost, germinates in light	leaf herb	'queen of herbs' fodder
butterflies	herb	pharmaceutics, fodder
sensitive to frost, P: March, SO: May	leaf	pharmaceutics, culinary use, fodder
resistant	leaf herb	tea, pharmaceutics fodder, large quantities; lactagogue

Sp = spring T = top cuttings
S = summer D = division
A = autumn

name (botanical name)	growing method	space required cm (row spacing)	sowing rate g/100 m² (TGW)	time
marjoram, pot (*Origanum majorana*)	P/SO a	20 × 20 (20–25)	P: 2–5 SO: 80–100	Sp
marjoram, wild (*Origanum vulgare*)	SO/T p	(25–30)	40–50	Sp
peppermint (*Mentha piperita*)	T/St p	40 × 20		Sp/A
rosemary (*Rosmarinus officinalis*)	P p		(0.8–1.3)	Sp
sage (*Salvia officinalis*)	P/SO/ T/D	30 × 30	P: 10–15 SO: 80–100	Sp
savory, summer (*Satureja hortensis*)	SO a	25 × 20 (25–30)	50–70	
savory, winter (*Satureja montana*)	SO p	(25–30)		Sp
thyme (*Thymus vulgaris*)	P/D/T p	20 × 20	P: 3–5 SO: 15–20	Sp
daisy family				
elecampane (*Inula helenium*)	P p	50 × 50	5	Sp
marigold, pot (*Calendula officinalis*)	SO a	40 × 40	200–400 (2.5–5)	Sp
mugwort (*Artemisia vulgaris*)	P/D p	40 × 50	(0.1)	Sp–A
southernwood, herb royal (*Artemisia abrotanum*)	P/T p	40 × 30	–	
tarragon (*Artemisia dracunculus*)	T/D p	30 × 40	–	Sp (A)
wormwood (*Artemisia absinthium*)	P p	40 × 40	(0.1)	Sp

p = perennial
a = annual
b = biennial

SO = sown in open
P = plant out
St = stolons

Table 36. continued.

9. ARABLE CROPS

comments	part used	use
2–3 yr cultivation, very aromatic, seed sterile	leaf herb	tea fodder
bee plant mixed crops	flowers herb	pharmaceutics, cosmetics fodder
	flowers herb	pharmaceutics
slightly sensitive to frost	leaf herb	culinary use, tea, pharmaceutics fodder
sensitive to frost, germinates in light	leaf herb	culinary use fodder
winter hardy, germinates in light	leaf	culinary use
bee plant germinates in light	flowers herb	culinary use, tea, pharmaceutics fodder
root mucilaginous	leaves root	culinary use pharmaceutics
	flower herb	tea, pharmaceutics, fodder
	leaf herb	culinary use fodder
seed difficult to obtain	herb	fodder, in small amounts
aroma poor if grown from seed	herb, at start of flowering	culinary use, fodder
very bitter	herb	tea, pharmaceutics, fodder in small amounts, liquor for fungi and pests

Sp = spring   T = top cuttings
S = summer   D = division
A = autumn

name (botanical name)	growing method	space required cm (row spacing)	sowing rate g/100 m² (TGW)	time
Other families				
borage (*Borago officinalis*)	SO a	(30)	70–80	Sp/S
burnet, salad (*Poterium sanguisorba/Sanguisorba minor*)	SO p	(30)	(5–10)	Sp
chives (*Allium schoenoprasum*)	SO/D p	20 × 20 (25)	20–25	Sp/S
comfrey (*Symphytum officinale*)	St p	50 × 50		Sp–S
fenugreek (*Trigonella foenum-graecum*)	p		150	
garlic (*Allium sativum*)	p	20 × 20	cloves	Sp/S
goat's rue (*Galega officinalis*)	SO/P p	60 × 60	20–30	Sp
horseradish (*Armoracia rusticana*)	St p	60 × 30 (60–70)	560 stolons	
rue (*Ruta graveolens*)	T/P p	25 × 30	10–12	Sp
scurvy grass (*Cochlearia officinalis*)	SO p		60–80	Sp
p = perennial		SO = sown in open		
a = annual		P = plant out		
b = biennial		St = stolons		

Table 36. continued.

9. ARABLE CROPS

comments	part used	use
germinates in light leaf		pharmaceutics, culinary use
needs thinning	herb	culinary use, fodder in large amounts
	herb	culinary use, liquor as fertilizer and for insect pests
	herb	fodder, medicinal for pigs, liquor as liquid manure
	herb	lactagogue, fodder
	cloves	culinary use, pharmaceutics
	herb	liquor for insect pests
bee plant, sensitive to frost, in hedges, waste ground	herb	lactagogue, fodder
	root	culinary use, pharmaceutics
	whole plant	plant around potato fields
	leaves	pharmaceutics, small amounts added to fodder
	herb	culinary use
bee plant	leaves	pharmaceutics, culinary use

Sp = spring   T = top cuttings
S = summer   D = division
A = autumn

203

Cultivation

The herbs grown in Central Europe originated in widely different climates, but are generally highly adaptable. Most of them like loose, humic soil in warm sites with moderate moisture, and either sunny positions (marjoram, sage) or semishade (lovage, tarragon). The soil is roughly dug in autumn. It is best to avoid manures that will force growth, as this reduces the scent and flavour. Well rotted farmyard manure or, even better, compost is preferable. Compost can be nitrogen enriched by adding hoof and horn chips. Dung liquor should not be applied during the period of active growth.

High standards are required for seed and seedlings. If home-grown material is not available, the best sources are other farms or market gardens well known to the farmer, or reliable specialist shops.

As Table 36 shows, many herbs are propagated from seed drilled directly in the open or with seedlings grown in seed beds and trays. With the last method, frost-sensitive annuals are planted out after mid-May and perennials by the end of August or early September at the latest. Seeds that germinate in the light should only have a light covering of soil. For those that germinate in the dark, drilling depth should not be more than three to five times the seed diameter.

Vegetative propagation is possible with perennial herbs. Top cuttings are taken in June, when they are firm enough, and put in slightly sandy soil or a mixture of peat and sand until roots develop. They are planted out in late summer or the next spring.

Propagation is also possible by dividing the rootstock of older, well-established plants; this is done in autumn or, with frost-sensitive subjects, in spring. The method also serves to rejuvenate stock and will result in more active growth.

Stolons of peppermint or horseradish are collected in autumn and planted out in spring. For details see the section on growing peppermint below.

Immersing roots for a few seconds in a root bath will help

seedlings, root and shoot cuttings. The bath is a thinnish paste made with two thirds clay and one third cow pats. Specially formulated products based on a variety of herbs and natural sulphur have also proved beneficial.

Cultivations needed are as follows:
- weed control (hoeing, hand weeding),
- regulation of leaf and root development (watering, cutting back, if necessary shading), and
- application of biodynamic field sprays: horn manure in spring and autumn and after every harvest, horn silica during early growth and prior to cutting.

Harvesting and processing

Relatively large quantities of leaves and shoots are harvested two or three times during the summer months. The final date should not be too late, to allow the plants to recover fully before winter sets in.

Aroma and active principles are usually at their highest level when the plant is coming into flower. Herbs that have reached this stage are cut or picked on a dry day in the morning after the dew has evaporated. The method of cutting will depend on the species, as plants react to this in widely differing ways. Careful observation of growth after harvest will provide the necessary information. Herbs may be dried to preserve them for winter. After removing all soiled or yellowed parts, they are well spread out on clean cloths and left to wilt in a darkened dry and airy room. The wilted plants are tied in bunches and hung to dry in a dry place. Drying is complete when the stalks break like glass. The leaves are stripped off and stored in closed containers (preferably not plastic).

Fruits and seeds, on the other hand, are collected in the early hours of the morning or under a cloudy sky, as fewer seeds are lost in this way. The time for harvesting is when the first fruits are turning brown.

Roots are dug in dry weather. They are cleaned to remove earth but not washed, as this would reduce their keeping quality.

Flowers are collected just before they open and on dry days. Repeated collections will be necessary, as flowers open successively. Spread in a shady place in the open or in an attic room.

Tables 36 and 37 show the many different plant parts that can be used. The carrot family in particular, but other families, too, include a wide variety of species that characteristically develop mainly one part. These find medicinal uses for humans and animals and help to balance biological functions.

Table 37 lists plants according to the lower (root), middle (stem/leaf) and upper (flower and seed/fruit) regions, whichever is the part used.

root	leaf/stem	flower/seed/fruit
carrot family		
angelica	celery	anise
carrot	angelica	caraway
celery	caraway	coriander
lovage	dill	dill
masterwort	fennel	fennel
parsley	lovage	
parsnip	masterwort	
	parsley	
mint family		
	basil	lavender
	hyssop	rosemary
	lemon balm	
	marjoram	
	peppermint	
	savory, summer	
	sage	
	thyme	
daisy family		
dandelion	mugwort	arnica
	golden rod	wild camomile
	southernwood	dandelion
	tansy	marigold, pot
	tarragon	sunflower petals
	wormwood	
other families		
comfrey	burnet, salad	elder
elecampane	chives	marsh mallow
garlic	comfrey	St John's wort
horseradish	fenugreek	
	goat's rue	
	horseradish	
	rue	
	St John's wort	

Table 37. Parts used of different herbs.

Growing peppermint

Peppermint is one of the most widely used highly aromatic medicinal plants. It provides a pleasant tea for family use, is used medicinally and is equally valuable in fodder mixtures. Dairy cows should only have a limited amount added to their rations, as the powerful essential oil can affect the taste of their milk.

The plants are grown on for two or three years, which puts them half way between annual and perennial crops, and they therefore have their own place both as a field crop and in the garden. After two or three years a new site is chosen, renewing the plant material at the same time. Propagation is always vegetative, using the stolons. Peppermint grows best in loose humic soils where the groundwater level is not too high; it likes a warm, sunny position. Rust is liable to develop if the soil is very acid or drainage is poor. Effective aeration and mulching improve plant growth and the quality of the essential oils.

Prepare a good bed in spring and give it an application of horn manure. Place the stolons about 5–10 cm (2–4 in) apart in rills about 5 cm (2 in) deep, spacing rows at 30–40 cm (12–16 in). Cover with earth or mature compost and firm well or roll. Spray horn silica as soon as vigorous shoots have appeared, and repeat two or three times after this, to increase the oil content. Cut with sickle or scythe as soon as the first inflorescences appear. Drying is best done in an airy shaded place.

Care must be taken not to make the second cut too late, as the aftermath comes into flower much more quickly. The other reason is that the plants need to recover fully before winter sets in.

Rust occurs especially if plants are left to flower for a long time, or if the second cut in autumn is late, so that the plants are weakened. Repeated spraying with horsetail when in full growth and later on the soil will prevent the fungus.

Mitcham is a specially recommended variety.

Herb fodder

Herb fodder is much easier to produce than plants for medicinal use. Expenditure is low, and less care is needed with harvesting, drying and processing.

Cut with a scythe or cutter bar. Leave to wilt in the sun for one to three hours before speading out well to dry in a dry, airy attic. The herbs will go mouldy if put in a damp, airless place. Leaves do not have to be stripped off, as the whole aerial part is used.

Further processing depends on the intended use:
- for teas, chop into 2–4 cm (1–1½ in) pieces;
- to add to crushed cereal grains and for the farm's own mineral mix, grind to powder (hammer mill).

In either case, operations should be carried out on dry days and only with the dried herbs.

Bibliography

Appel, J. 1979. *Unkrautregulierung ohne Herbizide.* Lebendige Erde. Darmstadt.
Darwin, C. 1980. *The Power of Movement in Plants.* John Murray. London.
Deutsch, A. 1972. *Pflanzenproduktion.* Leopold Stocker. Graz.
Elliot, R.H. 1898. *The Clifton Park System of Farming and Laying Down Land to Grass.* 6 ed. 1948. Faber & Faber. London.
Finck, A. 1976. *Pflanzenernährung in Stichworten.* Ferdinand Hirt. Kiel.
Fischbeck, G., Heylandt, K.-H., & Knauer, N. 1982. *Spezieller Pflanzenbau.* Ulmer. Stuttgart.
Goetz, A. & Konrad, J. 1978. *Landwirtschaftliches Lehrbuch.* Vol.1: *Pflanzenbau.* Ulmer. Stuttgart.
Grohmann, G. 1989. *The Plant.* 2 vols. Biodynamic Association of North America.
Hayward, H. E., and Spurr, W. B. 1943. Effects Literature of Osmotic Concentration of Substrate on the Entry of Water into Corn Roots. *Bot. Gaz.* 105, 152–164.
Hoffmann, M. 1980. Abflammtechnik. *KTBL Schrift.* 243. Darmstadt.
Huebner, R. 1955. *Der Same in der Landwirtschaft.* Neumann. Radebeul & Dresden.
Kiel, W. 1954. *Acker- und Pflanzenbau.* Deutscher Bauernverlag. Berlin.
Klapp, E. 1951. *Futterbau und Gruenlandnutzung.* Paul Parey. Hamburg/Berlin.
Klett, M. 2005. *Principles of Biodynamic Spray and Compost Preparations.* Floris. Edinburgh.
Knapp, R. 1971. *Einführung in die Pflanzensoziologie.* Ulmer. Stuttgart.
Koch, W. & Hurle, K. 1978. *Grundlagen der Unkrautbekämpfung.* Ulmer. Stuttgart.
Koepf, H.H. 2012. *Koepf's Practical Biodynamics.* Floris. Edinburgh.
Koepf, H.H., Pettersson, B.D., & Schaumann, W. 1976. *Bio-Dynamic Agriculture.* Anthroposophic. New York.
Kraus, C. 1914. Zur Kenntnis der Verbreitung der Wurzeln in Beständen von Rein- und Mischsaaten. *Frühling Landw. Zeitung,* 10. 337–412.
Künzel, M. 1949. *Über den Nutzen von Saatbädern. Neuaufbau Biologisch-Dynamischer Landbau 1945-1949.* Lebendige Erde. Stuttgart.
Larcher, W. 1984. *Ökologie der Pflanzen.* Ulmer. Stuttgart.
Masson, P. 2014. *A Biodynamic Manual: Practical Instructions for Farmers and Gardeners.* Floris. Edinburgh.

Padel, S. 1984. Anforderungen an Sorten im biologischen Getreidebau. IFOAM, No. 48, 1st Quarter.
Pfeiffer, E.E. 2011. *Pfeiffer's Introduction to Biodynamics.* Floris. Edinburgh.
Renius, W. 1978. Untersaaten im Gründüngungs-Zwischenfruchtbau. *Lebendige Erde.* 65-68. Darmstadt.
Ruebensam, E., & Rauhe, K. 1964. *Ackerbau.* Deutscher Landwirtschaftsverlag. Berlin.
Schechtner, G. 1969. *Aktuelle Fragen der Wirtschaftsdüngeranwendung auf dem Grünland.* Bundesvers.-Anstalt. Gumpenstein.
Schumacher, W. 1958. *Lehrbuch der Botanik, 2. Teil: Physiologie.* Gustav Fischer. Stuttgart.
Steiner, R. 1923. *Drei Perspektiven der Anthropsophie* (GA 225) Dornach 1990.
—. 1924. *The Agriculture Course.* Bio-Dynamic Agricultural Association. London 1974.
Thun, M. *The Maria Thun Biodynamic Calendar.* Floris. Edinburgh (annual).
Wiljams, W.R. 1958. *Das Trawolpolnajasystem der Landwirtschaft.* Berlin.

Index

abbreviations 8
air 13f
alfalfa *see* lucerne
algae, blue-green 14
ant hills 23
astronomy 46
autumn cultivation 97

bacteria, nitrogen-fixing 14f, 60, 79, 143
bacteria, soil 14
barley 44, 110, 115f
basalt rock dust 112
beans 144
bee plants 199, 201
bees 154
beetroot 136f
biodynamic preparations *see*
   camomile preparation,
   dandelion preparation,
   horn manure,
   nettle preparation,
   oak bark preparation,
   valerian preparation,
   yarrow preparation
birch pit concentrate 125, 130, 136
birdsfoot trefoil 144, 167f
— seed production 194
black grass *see* foxtail, slender
black medick 166
— seed production 193
blossom rape beetle 151
buckwheat 154

cabbage 141f
— caterpillar 142
— clubroot 142
— digestibility 141
— pest control 142
camomile 197
— preparation 80, 136, 148, 197
carbohydrates 11
carbon dioxide 12, 14f, 18
carbon/nitrogen ratio 50, 54, 61
carrot 119, 138–40
— horn manure 139f
— horn silica 139f
— manuring 138
— weed control 138
carrot fly 140
catch crops 54, 183f
— cropping sequences 138
— sowing data 186f
cereals 109
— as cover crops 159
— as feed 209
— disease control 117f
— drill fertilization 112
— horn silica 111
— smut 117f
— weed control 98f
— with peas 146
Clifton Park System 175
climate 19
clover 25, 40, 144, 162–64, 166f
— sickness 53f
—, Alsike 166
— — seed production 193
—, crimson 167
— — seed production 193
—, Egyptian 163

—, Persian 162
— — horn silica 162f
—, red 164
— — seed production 192
—, white 166f
— — seed production 193
Colorado beetle 122, 127f, 132
combine harvesting 98f
compost, on pasture 28
couch grass 52
cover crops 159f
cowpat preparation 151
crocus, autumn 36
crop plants 44
—, history of 44
crop rotation
—, arable and grassland 175
— improved three-field system 47
— polycrop 47f
— sidereal system 46
— two and three field system 47
cropping 44f
cropping sequences 46–72, 95–98, 158, 191f
— annual clovers 60
— catch crops 48, 55, 61–63
— cereals and grasses 52, 55–66
— compatibilities 52f, 65
— economic aspects 50f
— examples 64–72
— labour aspects 50f
— legumes 49–62
— leys 55, 58, 61
— linseed 60
— manuring 64
— nitrogen fixation 60
— oil crops 60
— percentage areas 48
— pest control 51f
— poppies 57
— potatoes 56–58
— principles 55
— pulses 60
— root residues 61

— row crops 60f
— soil-improving crops 49f, 58
— terminology 55
— weed control 51f, 96f
crumb structure 60f, 89–93, 109
cultivations, post drilling 91f

dandelion 206
— preparation 80
Darwin, Charles 17
disease control
— cereals 117f
— potato 128
drainage ditches, clearing 23
drainage, poor 22f
drill (implement) 90
drill fertilization, cereals 112
drought 75

ecology 91
eelworm
— beet 52
— stem and root 52
ergot 115

farm as organism 64
field beans 60, 146
flame weeding 102–4, 135, 139, 180
flax 152f
— horn manure 153
— seed bath 153
— seed bed 152
fodder crops 157–83
— beet 133
— — weed control 136
— seed bed 158
— seed quality 158f
— sowing data 181
fodder vs. green manure 63
foxtail, slender 94
— control 97
frost
— benefit for soil 64
— damage 33, 111
— protection 90

INDEX

fungi
— control 52
— *see also* disease control

geotropism 11
germinating power 82f
germination tests 82f
grasses
— drilling 160f
— key characteristics 172
grassland 21
— cultivation 22f
— herbage composition 28
— management 29–31
— manuring 23f, 28–30
— ploughing 38f
— renovation 38
— reseeding 40
— rolling 22f
— seed mixtures 40
— silage cuts 33
— weed control 35–38
grazing
— paddock 26f
— rotational 28
— spring and autumn 32f
— strip 26–30
green manure 134, 144, 146, 187
— vs. fodder value 63
gross margins 50

hay 135, 164, 170
— and grazing rotation 21, 26, 29
hedges 42
heliotropism 11
herbs 197–209
— data on growing and using 198–203
— for fodder 209
— harvesting and drying 205
— horn silica 205
— parts used 205
horn manure
— carrot 139f
— herbs 205

— linseed 153
— on fodder crops 163
— on grassland 24, 38, 42
— on pasture 28
— peppermint 208
— potato 131
— seed bed 90
— turnips 185
horn silica 75
— carrot 139f
— cereals 111
— herbs 205
— linseed 153
— maize 180
— marrowstem kale 182
— on cover crops 163
— on grassland 24
— on pasture 28
— Persian clover 162f
— potato 129
— sunflower 181
— turnips 185
horseradish 204
horsetail spray
— peppermint 208
— potato 132
horsetail, marsh 36

indicator plants 35, 94

land use 19f
Landsberg mixture 135, 138, 184, 190
legumes 31, 47–62, 143f
lentils 60, 144, 148
leys 169–79
— biodynamic sprays 177
— cutting 177
— duration 170f
— ploughing 177
— suitable mixtures 170f, 174f, 178
light 13f
linseed 60, 152f
— horn manure 153
— seed bath 153

215

liquid manure, application of 24, 185
lucerne 144, 160, 165, 174f, 192
— seed production 192
lupins 60, 121, 143, 148f
— seed production 149
— soil requirements 149

maize 60, 179f
— for silage 64, 120, 181, 190
— in cropping sequence 64, 68f
mangel *see* fodder beet
manure
— composted 64
— on grassland 28
marrowstem kale 182
— horn silica 182
meadows
— manuring 25
— timing first cut 25
— use 25f
— weed control 25f
melilot 144, 169
— seed production 194
mineral uptake, plants 109, 16f
mineralization 122
minerals 13, 15
molehills 23
mulching 94
mustard 46, 60, 154, 189
— with peas 145

nematodes 52
nettle
— as crop 155f
— as weed, control 38
— preparation 80
— tonic 142
nitrogen fixation 14, 60, 143

oak bark preparation 80
oats 110, 116f
oil crops 60, 109, 186
oil radish 60, 186f
— seed production 195
paddock grazing 27

pasture 22
— cleansing cuts 35
peas 60, 144–46
— frost 145
— seed bath 145
— seed bed 145
— soil requirements 144
— with cereals 146
— with mustard 145
peppermint 200, 208
— horn manure 208
— horsetail spray 208
— in fodder 208
perennial weeds, control 96
pest control
— cabbage 142
— cereals 118
phacelia 97, 134, 187f, 195
— seed production 195
photosynthesis 13f
phototropism 13
plants, nature of 11–18
plant breeding problems 73f
planting calendar 108
poisonous plants 35f
potassium 37
potato 56–58
— blackleg 124
— blight 120
— chitting method 124
— Colorado beetle 122, 132
— degeneration 120f
— diseases 128–32
— dynamic measures 128–32
— frost damage 120f
— haulms as mulch 23
— harvesting 132f
— horn silica 129
— horsetail spray 132
— in diet 120
— manuring 120f
— preceding crops 121
— preparing soil 121
— scab 120f
— seed bath 124

INDEX

— soil requirements 120
— stem canker 124
— true seed 121
— valerian spray 130
— weed control 127–128
preceding crop value 49, 56, 113f, 163
precipitation 20, 98
Preparation 500 *see* horn manure
Preparation 501 *see* horn silica
Preparation 502 *see* yarrow preparation
Preparation 503 *see* camomile preparation
Preparation 504 *see* nettle preparation
Preparation 505 *see* oak bark preparation
Preparation 506 *see* dandelion preparation
Preparation 507 *see* valerian preparation

rape 60, 150f, 188f
— seed bed 151, 188
rolling of fields 22f
rotational grazing 27
rye 105, 110, 113–15
— as green fodder 189
— breeding 114
— frost resistance 113f
— seed bed 113f
— winter 52
— with vetches 189f
ryegrass 41, 182f
— seed production 195

sainfoin 144, 168f
sales potential 50
sandy soil 15, 114
Schmidt rye 114
Seaweed, calcified 112
seed
— baths 75, 79–81
— —, carrots 138
— —, cereals 118
— —, fodder plants 163

— —, linseed 153
— —, peas 145
— —, potato 125f
— —, swedes 135
— bed 40, 89
— —, carrot 138
— —, cereals 111
— —, flax 152
— —, fodder beet 134
— —, fodder crops 158
— —, peas 145
— —, rape 151, 188
— —, rye 113f
— —, wheat 112f
— choice of variety 75
— cleanness 79f
— germinating power 82f
— production 75
— —, in-field selection 78
— —, lucerne 192, 196
— —, manuring 78
— —, negative mass selection 79
— —, oil radish 195
— —, phacelia 195
— —, positive mass selection 78
— —, preselection 78
— —, ryegrass 195
— —, sites 77
— —, vetches 195
— size, weight, shape 80, 84
— sorting by hand 78
— sowing times 76
— water content 83
serradella 144, 163, 183
— seed production 194
sidereal system 46
silage 135, 164, 170, 181, 190
slender foxtail, control 97
soil pH 121
soil water 21
soilage 63f, 170
soil-improving crops 130
sowing
— density 86
— depth 87

217

— distance between rows 88–90
soya bean 60, 144
spelt 113
steam sterilization 102
Steiner, Rudolf 17, 24, 76, 118, 120, 124
strip grazing 27
sugar beet 60, 133
sulphur 14
sunflower 60, 181
— horn silica 181
swedes 134f
— seed bath 136

thistles 37
— control 38, 52, 96f
Thun, Maria 108, 133
tilth 89f
transpiration 11–15
trees on grassland 126
trines (trigons) 128–32
true seed 124
turnips 185
— horn manure 185

valerian
— preparation 146f
— spray 130, 146, 153
vetches 60, 143
— seed production 195
— with rye 189f
vetchling 143

water, uptake by plants 14–17
watering animals in pasture 33f
weed control 35f, 51f, 91f
—, beet 136
—, carrot 138
—, cereals 98f
—, grassland 35–38
—, maize 102, 180
—, meadows 28
—, mechanical 98f
—, potato 127f
—, swedes 134f
—, thermal 102–4
weeds 91
— on grassland 36
— seeds 95–101
weevil, pea and bean 54
wheat 44, 51–57, 112f
— baking quality 113
— seed bed 112
wild oats, control 52, 97
wind protection 114f
woodland 19f, 43

yarrow preparation 80

The Maria Thun Biodynamic Calendar

Matthias Thun

This useful guide shows the optimum days for sowing, pruning and harvesting various plants and crops, as well as working with bees. It includes Thun's unique insights, which go above and beyond the standard information presented in some other lunar calendars. It is presented in colour with clear symbols and explanations.

The calendar includes a pullout wallchart that can be pinned up in a barn, shed or greenhouse as a handy quick reference.

This little paperback is essential for all organic gardeners... The original annual biodynamic planting guide.

– organicfood.co.uk

florisbooks.co.uk

The Biodynamic Farm

Developing a Holistic Organism

Karl-Ernst Osthaus

Large-scale agriculture tends to view a farm as a means for producing a certain amount of grain, milk or meat. This practical book argues instead for a holistic method of farming: the farm as a living organism. This is the principle of biodynamic farming.

The author, an experienced farmer, takes a down-to-earth approach. Based on an example farm of around 60 hectares, he recommends the ideal numbers of livestock: 12 cows, 4 horses, 6 pigs, 10 sheep and 120 hens. This mix is drawn from Osthaus's deep understanding of nature, animals, agriculture and the cosmos, and from his many years of personal experience as a biodynamic farmer and teacher. The result is a healthy, balanced and sustainable farm.

This is an invaluable book for anyone considering setting up a farm, or developing their existing farm with new biodynamic methods.

florisbooks.co.uk

A Biodynamic Manual

Practical Instructions for Farmers and Gardeners

Pierre Masson

For anyone already practicing, or turning to, biodynamic gardening and farming methods, numerous detailed questions arise, such as:

- How do you choose your seeds?
- What fertilisers should you use?
- Which natural products are most effective?

This manual, fully illustrated with explanatory diagrams and photographs, provides the answers. The book covers

- all aspects of making and using biodynamic preparations and composts
- managing the health of plants
- weed control
- parasite control
- issues around mixed cultivation
- animal care
- specialised crops and planting such as fruit trees and vines

This is an invaluable guide for all biodynamic growers to have to hand daily.

florisbooks.co.uk

Weeds and What They Tell Us

Ehrenfried Pfeiffer

This wonderful little book covers everything you need to know about the types of plants known as weeds. Ehrenfried Pfeiffer discusses the different varieties of weeds, how they grow and what they can tell us about soil health. The process of combatting weeds is discussed in principle as well as in practice, so that it can be applied to any situation.

First written in the 1950s, this is still one of the best overviews of the subject available.

florisbooks.co.uk

Pfeiffer's Introduction to Biodynamics

Ehrenfried Pfeiffer

Ehrenfried Pfeiffer was a pioneer of biodynamics in North America. This short but comprehensive book is a collection of three key articles introducing the concepts, principles and practice of the biodynamic method, as well as an overview of its early history.

The book also includes a short biography of Ehrenfried Pfeiffer by Herbert H. Koepf.

florisbooks.co.uk

You may also be interested in ...

The Biodynamic Orchard Book
Ehrenfried Pfeiffer and Michael Maltas

Koepf's Practical Biodynamics
Soil, Compost, Sprays and Food Quality
Herbert H. Koepf

What's So Special About Biodynamic Wine?
Thirty-five Questions and Answers for Wine Lovers
Antoine Lepetit de la Bigne

When Wine Tastes Best
A Biodynamic Calendar for Wine Drinkers
Matthias Thun

florisbooks.co.uk